International Marine Mammal Law

Nikolas Sellheim

International Marine Mammal Law

Springer

Nikolas Sellheim
Helsinki Institute of Sustainability Science (HELSUS)
University of Helsinki
Helsinki, Finland

Scott Polar Research Institute
University of Cambridge
Cambridge, UK

Polar Research and Policy Initiative (PRPI)
London, UK

ISBN 978-3-030-35267-7 ISBN 978-3-030-35268-4 (eBook)
https://doi.org/10.1007/978-3-030-35268-4

This Springer imprint is published by the registered company Springer Nature Switzerland AG.
The registered company address is: Gewerbestrasse 11, 6330 Cham, Switzerland

Dedicated to my mother

Foreword

The book *International Marine Mammal Law* by Nikolas Sellheim is a commendable undertaking. It seeks to address a range of misconceptions, myths, ill-founded allegations, and occasional fake news on marine mammals, the threats they currently face, the worldwide status of their populations, the people and countries hunting them, and the rules of international law relating to their conservation and sustainable use. This serves a twofold purpose. The first essentially amounts to fact-checking, which is nowadays increasingly resorted to in response to fake news—whether manmade or by robots.

The second purpose is to provide insights into the multiple perspectives and dimensions that have made conservation and sustainable use of marine mammals such a controversial issue. Prominent among these are the fundamental differences in societal values that exist in individual states as well as the foundational principles of international law: the sovereign equality of states and the consensual nature of international law. The need for prior consent means that states cannot be bound against their will. This second purpose of the book is ultimately aimed at fostering mutual understanding and respect among citizens and states. Ideally, children will be taught to be careful when judging others. That thinking in terms of good and bad, or superior and inferior, will only drive people apart and not bring them together. That it is more constructive to acknowledge that, often, things are just different.

This book was written during the period when Japan announced its decision to withdraw from the International Convention for the Regulation of Whaling—and thereby from the International Whaling Commission (IWC)—and to commence whaling in its own maritime zones, rather than far way, such as off Antarctica. The actual start of this whaling on 1 July 2019 heralds a new chapter in international marine mammal law in the twenty-first century, with large-scale commercial whaling now occurring formally outside the auspices of the IWC. How long this unregulated whaling will continue depends on a range of factors, including the evolving opinions of citizens on hunting and consumption of animals, and their welfare; the worldwide status of whale populations; global, regional, or domestic food shortages; and, last but not least, the willingness and ability of media worldwide to report on marine mammals in a well-informed, fact-based, and neutral manner.

Netherlands Institute for the Law of the Sea (NILOS) Erik J. Molenaar
Utrecht, The Netherlands

Utrecht Centre for Water
Oceans and Sustainability Law (UCWOSL)
Utrecht University
Utrecht, The Netherlands

Norwegian Centre for the Law of the Sea (NCLOS)
UiT The Arctic University of Norway
Tromsø, Norway
July 2019

Preface

This book is the outcome of several years working on marine mammal cultures and regulation. The motivation to write this book stemmed from the fact that in many discussions surrounding marine mammals, marine mammal conservation, or even cultures and communities engaged in hunting marine mammals, many misconceptions have risen to the surface. First and foremost, claims that whales are universally protected, that polar bears are dying out and killing them is illegal, or that seal hunting is illegal in the European Union are difficult to rebut in social settings. I therefore decided that a rather short introduction to 'international marine mammal law'—a term which as such does not exist—is quite necessary.

I therefore embarked on a journey and decided to write a not-so-scholarly book that is also understandable for the lay reader not familiar with the topic. As a scholar, I am used to write in academic jargon, to use big words, and to take for granted that specific concepts or contexts are known. In this book, I have tried to avoid this in order to make it easily understandable for all readers and to make the complexities of international law pertaining to marine mammals accessible. In order to do so, I have also used my own experiences and included my own research results—either already published or still in the process of being written down in a scholarly manner.

The outcome of this endeavour is this textbook, which serves as a point of departure for further studies on international marine mammal law and which can be used as course material in universities or even as a basic work for individuals or organisations working on marine mammal related issues. Since I do not expect that the readers of this book are fully acquainted with international law, I included some basics of international law as well. By the same token, since I do not expect the reader to be familiar with the mysterious group of animals commonly known as marine mammals, I also included a chapter on some marine mammal species. A comprehensive volume for students, for lay persons, or for marine mammal professionals working in other areas than law is the outcome that you are now holding in your hands.

The information contained in this book stems from my activities as a researcher, but are compiled and considered in a way suitable for a non-academic audience. Some parts summarise my doctoral research at the University of Lapland, Finland, where I analysed the EU Regime on Trade in Seal Products. Moreover, it includes research results from my first postdoctoral project entitled 'Development of

International Law on the Conservation of Marine Mammals' (2017–2018), carried out at the Polar Cooperation Research Center, Kobe University, Japan. The project was funded by the Japan Society for the Promotion of Science. As such, the book has emerged as a 'side product' of my postdoctoral project 'Livelihoods, Cultures and Local Communities in International Conservation Law' (2018–2020), funded by the Maj and Tor Nessling Foundation, carried out at the Helsinki Institute for Sustainability Science (HELSUS), University of Helsinki.

The international legal regimes dealing with marine mammals are subject to swift changes and fast developments. Upon the time this book will be published, I am sure some of the information will already be dated. However, while that may be so, I hope I have succeeded in providing some insights into the complex interplay between marine mammals and international law.

Lahti, Finland
Summer 2019

Nikolas Sellheim

Acknowledgements

When writing a book on whatever topic, it is almost impossible to say 'thank you' to all of the people who have been involved in the process leading to publication. For this book, I therefore focus on those dear people who have in one way or another contributed either through their knowledge or through their support in other ways throughout the years that have finally led to me writing this book.

First of all, without the funding of the Maj and Tor Nessling Foundation, the book would not have been possible, so I thank the foundation for its support of my research. But carrying out research always needs guidance and support from other scholars. I thank Professor Timo Koivurova and Professor Florian Stammler, Professor Lassi Heininen (all University of Lapland), Professor Akiho Shibata (Kobe University), and Professor Reetta Toivanen (University of Helsinki) for their support in all of my research endeavours.

Fieldwork in Newfoundland was made possible by Dion Dakins (Carino Ltd.), Dion Randell and the crew of *Steph & Tahn*, and the Canadian Sealers Association. Fieldwork in Taiji was made possible by Chikao Komura (Japan Small-Type Coastal Whaling Association), Yoshiko Saito, and Professor Akira Saito (Kobe University). In both cases I also thank the people of Woodstock, Twillingate, and Taiji for their willingness to spend their time to talk to me and to introduce me to their communities. I also thank Professor Joji Morishita (Tokyo University of Marine Science and Technology), Eugene Lapointe (IWMC World Conservation Trust), Dr Geneviève Desportes (North Atlantic Marine Mammal Commission), Professor Jun Akamine (Hitotsubashi University), Magni Arge (Faroese Delegation to the Danish Parliament), and Nauja Bianco (Isuma Consulting) for their support. I would like to express my gratitude to Lotta Manninen (WWF Finland), Sheryl Fink (International Fund for Animal Welfare), Devon Joseph (IWC Commissioner for Antigua and Barbuda), and Herman Oosthuizen (IWC Commissioner for South Africa) for their support. I also thank Dwayne Ryan Menezes (Polar Cooperation Research Initiative), the staff of the International Secretariat of the University of the Arctic, and the staff of the Arctic Centre, University of Lapland.

I express my sincere gratitude to Erik Molenaar (Utrecht University and UiT The Arctic University of Norway) for his Foreword and for his comments on the draft of the book as well as to Alan Turner (Lahti) for proofreading the book and for providing me with helpful comments. A special 'thank-you' also goes to Brigitte

Reschke at Springer for her help and guidance. While they all may have helped me completing this book, it is obviously clear that any errors or omissions are my own. On a more personal level, my deepest appreciation and friendship goes out to Helena Nelding-Kosuge and Ryusuke Kosuge as well as to Yukiko and Naonori Kishi, Nicola Künkel, Ian Kattein, Marzia Scopelliti, Leilei Zou, Yulia Zaika, Shadi Sakran, Nina and Teemu Kirjonen, Peter and Maike Sellheim, Doris Sellheim, Yukino Thompson and Konstantin Sellheim, Katharina Sellheim and Christian Zimmer-Sellheim, Helena and Matti Seppänen, Anu and Mikko Kähkönen, Daniel Claassen, Christina Witt, Ulf Scheel, Maria Mattelmäki and Yuky Ryang.

Last but not least, the love of my wife and children is a love only the happiest of men can receive.

Thank you all.

Contents

List of Acronyms

ABA	Arctic Biodiversity Assessment
ABWC	Alaska Beluga Whale Committee
ACC	Alaska Commercial Company
ACCOBAMS	Agreement on the Conservation of Cetaceans in the Black Sea, Mediterranean Sea and contiguous Atlantic Area
ACPB	Agreement on the Conservation of Polar Bears
AEPS	Arctic Environmental Protection Strategy
AIBWC	Alaska and Inuvialuit Beluga Whale Committee
AIDCP	Agreement on the International Dolphin Conservation Program
ASCOBANS	Agreement on the Conservation of Small Cetaceans of the Baltic, North East Atlantic, Irish and North Seas
ASW	Aboriginal Subsistence Whaling
ATCM	Antarctic Treaty Consultative Meeting
ATCP	Antarctic Treaty Consultative Parties
ATS	Antarctic Treaty System
BWU	Blue Whale Unit
CAMLR	Convention for the Conservation of Antarctic Marine Living Resources
CBD	Convention on Biological Diversity
CCAS	Convention for the Conservation of Antarctic Seals
CITES	Convention on International Trade in Endangered Species of Wild Fauna and Flora
CLA	Catch Limit Algorithm
CLCS	Commission on the Limits of the Continental Shelf
CMS	Convention on Migratory Species of Wild Animals
CoE	Council of Europe
CoP	Conference of the Parties
DFO	Department of Fisheries and Oceans
EBSA	Ecologically or Biologically Significant Marine Areas
EEC	European Economic Community
EEZ	Exclusive Economic Zone
EFSA	European Food Safety Authority
EU	European Union

FAO	Food and Agriculture Organization of the United Nations
GATT	General Agreement on Tariffs and Trade
HELCOM	Baltic Marine Environment Protection Commission; Helsinki Commission
HSI	Humane Society International
HSUS	Humane Society of the United States
IATTC	Inter-American Tropical Tuna Commission
ICES	International Council for the Exploration of the Sea
ICJ	International Court of Justice
ICNAF	International Commission for the Northwest Atlantic Fisheries
ICR	Institute of Cetacean Research
ICRW	International Convention for the Regulation of Whaling
IFA	Inuvialuit Final Agreement
IFAW	International Fund for Animal Welfare
IGC	Inuvialuit Game Council
IMCO	Committee on Internal Market and Consumer Protection
IMMA	Important Marine Mammal Area
IMMC	International Marine Mammal Commission
ITK	Inuit Tapiriit Kanatami
IUCN	International Union for Conservation of Nature
IVWG	Independent Veterinarians' Working Group
IWC	International Whaling Commission
MEP	Member of the European Parliament
MMPA	Marine Mammal Protection Act
MMR	Marine Mammal Regulations
MoU	Memorandum of Understanding
MPA	Marine protected area
MSY	Maximum sustainable yield
NAFO	Northwest Atlantic Fisheries Organization
NAMMCO	North Atlantic Marine Mammal Commission
NMP	New Management Procedure
NSB	North Slope Borough
PBSG	Polar Bear Specialist Group
PETA	People for the Ethical Treatment of Animals
RFMO	Regional Fisheries Management Organisation
RMP	Revised Management Procedure
RMS	Revised Management Scheme
SOWS	Southern Ocean Whale Sanctuary
SSCS	Sea Shepherd Conservation Society
STCW	Small-type coastal whaling
TAC	Total allowable catch
TFEU	Treaty on the Functioning of the European Union
UNCCD	United Nations Convention to Combat Desertification
UNCED	United Nations Conference on Environment and Development
UNCHE	United Nations Conference on the Human Environment

UNCLOS	United Nations Convention on the Law of the Sea
UNEP	United Nations Environment Programme
UNFCCC	United Nations Framework Convention on Climate Change
WAZA	World Association of Zoos and Aquariums
WCED	World Commission on Environment and Development
WGMME	ICES Working Group on Marine Mammal Ecology
WTO	World Trade Organization

List of Figures

List of Tables

List of Treaties and Other Legal Instruments

Agreement between the Government of Canada and the Government of Norway on Sealing and the Conservation of the Seal Stocks in the Northwest Atlantic, 15 July 1971 (870 UNTS 85)
Agreement for the Conservation of Dolphins, 21 Apr 1992
Agreement for the Creation of a Sanctuary for Marine Mammals, 25 Nov 1999
Agreement on Cooperation in Research, Conservation and Management of Marine Mammals in the North Atlantic, 9 Apr 1992
Agreement on Measures for Regulating the Catch and Conserving Stocks of Seals in the Northeastern Part of the Atlantic Ocean, 22 November 1957 (1958 UNTS 280)
Agreement on the Conservation and Management of the Alaska-Chukotka Polar Bear Population, 16 Oct 2000
Agreement on the Conservation of Cetaceans in the Black Sea, Mediterranean Sea and contiguous Atlantic area, 24 Nov 1996 (2183 UNTS 203)
Agreement on the Conservation of Polar Bears, 15 Nov 1973 (13 ILM 13)
Agreement on the Conservation of Seals in the Wadden Sea, 16 Oct 1990
Agreement on the Conservation of Small Cetaceans of the Baltic and North Seas, 17 March 1992 (1772 UNTS 217)
Agreement on the International Dolphin Conservation Program, 21 May 1998
An Act to provide for the orderly Termination of Federal management of the Pribilof Islands, Alaska. (Public Law 98-129), 14 October 1983
Antarctic Treaty, 1 Dec 1959 (402 UNTS 71)
Award Between The United States and the United Kingdom Relating to the Rights of Jurisdiction of United States in the Bering's Sea and the Preservation Of Fur Seals, 15 Aug 1893
Consolidated version of the Treaty on the Functioning of the European Union, OJ C 326, 26.10.2012, pp. 47–390
Constitution of the State of Alaska, 24 Apr 1956
Convention for Co-operation in the Protection and Development of the Marine and Coastal Environment of the West and Central African Region, 23 Mar 1981 (20 ILM 746)
Convention for the Conservation of Antarctic Seals (CCAS), 1 June 1972 (11 ILM 251)

Convention for the Conservation of Antarctic Seals, 1 June 1972 (11 ILM 251)
Convention for the Preservation of the Fur Seal and Sea Otter in the North Pacific Ocean and Bering Sea, 6 Nov 1897
Convention for the Prevention of Marine Pollution by Dumping from Ships and Aircraft (Oslo Convention), 15 Feb 1972 (932 UNTS 3)
Convention for the Prevention of Marine Pollution from Land-Based Sources (Paris Convention), 4 June 1975 (1546 UNTS 119)
Convention for the Protection of the Marine Environment of the Northeast Atlantic, 22 Sept 1992 (2354 UNTS 67)
Convention for the Protection of the Mediterranean Sea Against Pollution, 16 Feb 1976 (1102 UNTS 27)
Convention for the Regulation of Whaling of 24 September 1931 (155 LNTS 349)
Convention on Biological Diversity, 5 Jun 1992 (1760 UNTS 79)
Convention on International Trade in Endangered Species of Wild Fauna and Flora, 3 Mar 1973 (993 UNTS 243)
Convention on the Conservation of Antarctic Marine Living Resources, 20 May 1980 (1329 UNTS 48)
Convention on the Conservation of Antarctic Marine Living Resources, 20 May 1980 (1329 UNTS 48)
Convention on the Conservation of European Wildlife and Habitats, 19 Sept 1979 (1284 UNTS 209)
Convention on the Conservation of Migratory Species of Wild Animals, 23 June 1979 (1651 UNTS 333)
Convention Relative to the Preservation of Fauna and Flora in their Natural State, 8 Nov 1933 (1 LNTS 172)
Convention Respecting Measures for the Preservation and Protection of the Fur Seals in the North Pacific Ocean (Fur Seal Convention), 7 July 1911
Council Regulation 338/97/EC on the protection of species of wild fauna and flora by regulating trade therein, OJ L 61, 3.3.1997, p. 1
Directive 92/43/EEC on the Conservation of Natural Habitats and of Wild fauna and Flora, OJ L 206, 22.7.1992, p. 7
General Agreement on Tariffs and Trade, 14 Apr 1994 (55 UNTS 194)
Interim Convention on Conservation of North Pacific Fur Seals, 9 Feb 1957
International Agreement for the Regulation of Whaling of 8 June 1937 (190 LNTS 79)
International Convention for the Northwest Atlantic Fisheries, 8 Feb 1949 (157 UNTS 158)
International Convention for the Prevention of Pollution of the Sea by Oil, 12 May 1954
International Convention for the Regulation of Whaling, 2 December 1946 (161 UNTS 72)
Inuvialuit Final Agreement, 5 Jun 1984
Inuvialuit Inupiat Beaufort Sea Beluga Whale Agreement, 3 Mar 2000
Marine Mammal Regulations, SOR/93-56, 4 Feb 1993

Memorandum of Understanding between Environment and Climate Change Canada and the United States Department of the Interior for the Conservation and Management of Shared Polar Bear Populations, 2008.

Memorandum of Understanding between the Government of Canada, the Government of Nunavut, and the Government of Greenland for the Conservation and Management of Polar Bear Populations, 31 Oct 2009

Memorandum of Understanding Concerning the Conservation of the Manatee and Small Cetaceans of Western Africa and Macaronesia, 8 Oct 2008

Memorandum of Understanding for the Conservation of Cetaceans and their Habitats in the Pacific Islands Region and Action Plan, UNEP/CMS/PIC-1/Inf/3, 15 Sept 2006

Peace Treaty between the Republic of Finland and the Russian Socialist Federal Soviet Republic, 14 October 1920

Polar Bear Management Agreement for the Southern Beaufort Sea, 29 Jan 1988

Protocol Amending The 1937 International Agreement For The Regulation Of Whaling of 24 June 1938 (196 LNTS 131)

Protocol Amending The 1937 International Agreement For The Regulation Of Whaling of 26 November 1945 (11 UNTS 45)

Protocol on Environmental Protection to the Antarctic Treaty, 4 Oct 1991 (30 ILM 1455)

Regulation (EC) No 1007/2009 of the European Parliament and of the Council of 16 September 2009 on Trade in Seal Products, OJ L 286, 31.10.2009, pp. 36–39

Regulations for Maritime Hunting Operations in the Waters of the South Pacific, 18 Aug 1952

Regulations Respecting the Protection of Seals, Statutory Order Regulations 64/443(Order in Council P.C. 1964/1663), 29 October 1964.

Suomen ja Venäjän sopimus kalastuksen ja hylkeenpyynnin harjoittamisesta Laatokalla [Finnish and Russian Agreement on Fishing and Sealing Operations in Lake Ladoga], 28 October 1922

Suomen ja Venäjän välinen sopimus kalastuksen ja hylkeenpyynnin harjoittamisesta molempien maiden aluevesillä Pohjoisella Jäämerellä [Agreement between Finland and Russia on fishing and sealing operations in the waters of both countries in the Northern Arctic Ocean], 21 October 1922

United Nations Agreement for the Implementation of the Provisions of the United Nations Convention on the Law of the Sea of 10 December 1982 relating to the Conservation and Management of Straddling Fish Stocks and Highly Migratory Fish Stocks, 4 Aug 1995 (2167 UNTS 88)

United Nations Convention on the Law of the Sea, 10 Dec 1982 (1833 UNTS 3)

United Nations Convention to Combat Desertification in Countries Experiencing Serious Drought and/or Desertification, Particularly in Africa, 17 June 1994 (1954 UNTS 3)

United Nations Declaration on the Rights of Indigenous Peoples, 13 Sept 2007

United Nations Framework Convention on Climate Change, 9 May 1992 (1771 UNTS 107)

Why an Introduction to International Marine Mammal Law? 1

1.1 Introduction

I was born in Germany in 1980—a year when the *Voyager* probe proved the existence of the Saturn moon Janus and when the US Olympic hockey team defeated the Soviets in the semifinals of the Winter Olympics. Of course, many other things happened, but all I was interested in was eating, drinking and sleeping. I was still a toddler when in 1982 the International Whaling Commission (IWC) decided to put a halt on the commercial hunt for whales. This so-called 'moratorium' on commercial whaling is probably one of the most far-reaching decisions the IWC has ever made. The reason is rather simple: while it is supported by the majority of the members of the IWC, it is far from ever having been a unanimous decision. Until today, the moratorium constitutes one of the—if not *the*—most contentious issues within the Commission. I have come across many statements in the German- and English-language media which claim that 'the world has made whaling illegal' and therefore that 'whales are protected under international law.' Also in discussions that are ongoing in social circles—actual or electronic—countries like Japan, which has always pushed for a resumption of sustainable commercial whaling, is often portrayed as the outlaw, the free-rider that ignores the world's wish to end the lethal and commercial use of whales.

Unfortunately, the world is not that simple. And also the legal world surrounding marine mammals—to which whales belong—is not that simple. Because when stating that 'the world has made whaling illegal' and that 'whales are protected', this significantly oversimplifies, and arguably neglects, at least three important legal facts: first, it is not 'the world' which has made whaling illegal, but rather the *majority* of the member states of the IWC, making the commercial hunt for whales illegal merely for its member states, counting 89 at the time of writing. In other words, from the perspective of international law, non-IWC-members, comprising all other 106 countries of this world, can in principle still hunt whales for commercial gain, including Japan, whose membership ended on 20 June 2019. Moreover, some IWC Members are not bound by the moratorium (e.g. Norway) as they exercised

N. Sellheim, *International Marine Mammal Law*,
https://doi.org/10.1007/978-3-030-35268-4_1

their right to opt-out. Second, under IWC rules 'whaling' as such has not been placed under a moratorium, but merely commercial whaling. As we will see, whaling for scientific purposes as well as whaling carried out by indigenous populations can still be pursued. Third, it is not 'whales' as an all-encompassing super species, which the IWC has protected, but rather those *whale species* falling under its purview. We will get to know those in Chap. 4. This means, that other whale species, those that are not dealt with by the International Whaling Commission, are by no means protected by the moratorium. These few lines, with rather little effort already demonstrate the crux of the issue: while there seems to be a social consensus on specific issues relating to marine mammals in some parts of the world, the legal circumstances are significantly more nuanced. Especially those not familiar with these legal aspects enter a slippery slope when arguing about legal issues surrounding marine mammals. Seemingly simple contexts are indeed complex legal situations.

This book aims to shed light on these complex situations by providing an account of the legal and cultural environments of international marine mammal law. You, the reader, will be able to familiarise yourself with many of the species in question and with the cultural elements associated with marine mammals in different parts of the world. I have researched the issues of marine mammals in Arctic and non-Arctic contexts for several years now and have conducted field research in the seal hunts in Canada and in the whale and dolphin hunts in Japan. Moreover, I have attended meetings of the International Whaling Commission as an observer. Even though I hold a doctorate in law, my research has always been linked with empirical data. This book is no exception. You will not only find law-paragraphs here, but instead the book is systematically infused with data collected from all over the world. After all, marine mammals are species that can be found all over the planet. Reading this book might contribute to you, the reader, appreciate the species themselves and the people that have interacted with them for centuries and even millennia. It is my view that this introduction to the world's legal regimes that regulate their protection, ensure their conservation and enable their utilisation appears to be a necessary step for us all to understand these creatures and their role in human societies a bit better.

1.2 The Structure of the Book

The following chapters provide you with some rather diverse information on the legal environment concerning marine mammals. Since it is my intention to take the issue beyond the confines of the academic world, I have abstained from providing references to each claim I make and therefore the book is rather a textbook than a 'traditional' scholarly volume. The reason for this is that this book is intended for a broader audience and an uninterrupted flow of reading. A 'massgrave of footnotes'—a quite fitting term by my former professor Jaakko Husa (University of Helsinki)—will not be part of this book. Instead, I have maintained an uninterrupted narrative. That being said, at the end of each chapter you will find the References-section in which I have provided literature that I have drawn from or that is relevant for the purposes of the book. Unless I cite an author directly, references in footnotes or elsewhere are therefore comparably absent in this book.

As the title of the book suggests, the main focus lies on the *legal* aspect concerning marine mammals. But it is my view that without understanding the biophysical and cultural dimensions of marine mammals as well as the interactions between marine mammals and human societies, the understanding of the current international legal framework is incomplete. This view was manifested by my past and ongoing research on marine mammal issues. It is my conviction that in order to understand the legal framework, one must understand the cultural and social frameworks, too. Of course, I'm far from being the first and only one having realised this, but I have yet to come across a strategic, socio-legal assessment of international law pertaining to marine mammals. Indeed, I would even go so far as to claim that a cultural forum concerning marine mammals is key for the IWC to overcome the deep divisions that have marked its existence since it has become clear that a significant number of states would not support a lifting of the 1982 moratorium.

In order to contribute to a better understanding of marine mammal species and marine mammals in different cultures, this book takes these elements explicitly into account. To this end, Chap. 2 provides some basic information on marine mammals. I have thus compiled information from leading scholars of marine mammal biology and physiology in order to provide information that make these stunning creatures better understandable. The underlying question of this chapter is therefore: what are marine mammals? We thus learn about the life history, their conservation statuses and what some of their biggest threats are.

With this in mind we turn to Chap. 3, which briefly outlines the social and cultural relationships between humans and marine mammals. Of course, a small book like the present cannot do justice to the myriad of facets of socio-cultural interaction all over the world, so the focus is shifted to some key regions, such as the Arctic, the North Atlantic, Eastern Canada or a country like Japan. In this chapter we will learn how perceptions towards marine mammals have changed—or remained—over time and how in everyday life marine mammals occupy a social role. While drawing from existing literature, I have incorporated data from my own research all over the world into this chapter. You, the reader, will thus be happy to learn that you will also find quite a few pictures here.

Starting with Chap. 4 we turn to the main core of the book. Here we will begin to find answers to the question 'what is international marine mammal law'? In fact, the singular use of the term 'law' is probably somewhat deceptive since there is no coherent body of international law that deals with marine mammals. Instead, we find scattered legal instruments, frameworks and bodies that in one way or the other deal with marine mammals. In this chapter we get to know international environmental law regimes as well as marine mammal-specific regimes of both a global and a regional character. We get to know cross-species regimes and institutions in which marine mammals are either explicitly or implicitly considered. Examples are the United Nations Convention on the Law of the Sea (UNCLOS), the Convention on International Trade in Endangered Species (CITES) and the North Atlantic Marine Mammal Commission (NAMMCO).

Chapter 5 deals with whales and considers all relevant international agreements that make exclusive reference to whales. At the centre of this part stands the International Convention for the Regulation of Whaling (ICRW) and the International Whaling Commission established by it. Apart from merely describing the convention, the part delves into the underlying problems of the IWC and what makes the current situation so contentious. Since the IWC merely deals with 'great whales' and leaves out small cetaceans, this chapter furthermore includes some pages on international cooperation regarding whale species outside the regulatory framework of the IWC.

Chapter 6 is concerned with seals and introduces the current international legal frameworks for the management and conservation of seals. Given the absence of a global agreement on the management of seals, this part takes the European Union's ban on trade in seal products under closer scrutiny and what implication this has had for the overall trade in seal products worldwide.

Chapter 7 looks at polar bears and specifically the Agreement on the Conservation of Polar Bears. We also get to know some bilateral agreements that serve the protection of polar bears. Moreover, the importance of the polar bear also within a CITES context is being discussed.

Chapter 8 delves into the question of *why* specific regimes pertaining to marine mammals are what they are. Taking into account all previous chapters it aims to explain the stances of particular countries, for instance within the International Whaling Commission or elsewhere. The chapter thus links comparative legal research—albeit understandable for lay readers—with political and cultural considerations and makes the prevailing legal regimes relevant for marine mammals explainable. The chapter serves consequently somewhat as a summary and discussion of the preceding chapters and is relevant in order to fully comprehend international marine mammal law(s).

Finally, Chap. 9 summarises all preceding chapters and asks where we currently stand concerning international marine mammal law. It provides an overview of the key challenges and key achievements of the current legal regimes and furthermore presents several recommendations on how to improve the current different systems, particularly discussing ways to reconcile diverging views on marine mammal conservation and utilisation.

1.3 Research on Marine Mammal Law

International law surrounding marine mammals has been the subject of many scholarly volumes, particularly in the English language. In fact, I have yet to come across scholarly or non-scholarly books in my native language German or in the language of my country of residence—Finland—that exclusively deal with the interaction of marine mammals and international law. Generally speaking, the main body of literature and research has focused on the International Whaling Commission and the International Convention for the Regulation of Whaling. Of particular interest have been the way the IWC works and what factors have influenced its current state, how whales are legally treated, and how the ICRW is implemented. Seals have first and foremost been considered in the context of the EU's ban on trade in seal products and within the World Trade Organization's (WTO) dispute settlement procedure.[1] Polar bears, by and large, are not quite as prominent in international legal scholarship. While there are notable scholars such as Professor Nigel Bankes at the University of Calgary who has looked at legal implications of co-management and trade law, the legal framework for polar bears is not quite as prominent in the international discourse as one might assume. Of course, although polar bears do appear in the contexts of international environmental law, climate change law and Arctic law, they are often part of a larger narrative but not necessarily an exclusive focal point therein.

To those unfamiliar with legal research, you might be unaware of the fact that legal scholars often go beyond the study of clauses, paragraphs and articles—meaning, they also take other issues into account than the actual legal text. Particularly in the twentieth century the scholarship of socio-legal research has evolved which includes social legal studies, anthropological legal studies or legal anthropology. In other words, scholars have started to look at the social life of law(s) and what role societies, people and peoples play in it. This book is no exception and makes the 'social legal life' of marine mammals better comprehensible. While social and cultural issues, paired with some rudimentary information on certain species are indeed part of many studies, in this book I intend to go further and thus provide a greater background. Of course it is up to you, the reader, to determine whether I have succeeded in doing so.

1.4 The Basics of International Environmental Law

Before we delve into the international marine mammal law itself, I consider it useful to write a few words about the nature of public international law and its subfield, international environmental law.

First of all, the term *public* international law refers to the law that has been agreed upon between nation states. That means that nation states are the subjects of public international law: they make the decisions, they are the ones responsible for compliance and implementation. This stands in contrast to *private* international law, which concerns the rights an obligations of citizens in different countries. Since this is not relevant for our purposes, when I refer to 'international law', I refer to *public* international law, the law of states.

Generally speaking, international law is guided by the principle of consensus, meaning that every internationally agreed legal text is the outcome of long negotiations in which each word is carefully considered so that every single state, which participates in the negotiations is content. Of course, there are different ways to steer negotiations and the final outcome, but for our purposes this should be

[1] Sellheim (2018).

sufficient. Yet, let me also mention the so-called 'free riders'. Since there are only very limited ways to enforce international agreements, states are also able *not* to follow specific provisions. That means even though they might be party to a specific treaty, they still don't comply with it. All international legally-binding treaties are guided by the principle of *pacta tertiis nec nocent nec prosunt*. This complicated Latin phrase translates into: a treaty bind the parties and only the parties. That means that every state that is not party to a specific treaty is not legally bound to it.

With this *very* short introduction to public international law may give you an idea of the basic principles that are relevant for this book. Let us now turn to international environmental law to which international marine mammal law first and foremost belongs. Of course, since marine mammals live in the sea, it is also the law of the sea which is a crucial element of international marine mammal law, but in the interest of scope, I will abstain from an introduction to the law of the sea since its main issues of relevance will be dealt with in Sect. 4.2.3 anyway. But I would recommend Churchill and Lowe's excellent treatise on the law of the sea for more information.[2] Let me also make clear that the branch of 'international marine mammal law' does not exist as such. It is a term which I use throughout this book in order to describe the bodies of law (legally binding or legally non-binding) which in one way or the other deal with marine mammals. By and large, as said above, this branch comprises legal regimes from international environmental law. Others would use the term 'international conservation law'[3] to describe those regimes that deal with the conservation and sustainable use of biodiversity. For those interested in international environmental law in general, I can highly recommend Timo Koivurova's *Introduction to International Environmental Law*.[4]

Briefly summarised, international environmental law in its modern form saw its emergence in the 1960s with the so-called 'global awakening' on environmental issues. Key for this development was the publication of the book *Silent Spring* by Rachael Carson in 1962.[5] This book deals with the interrelation between environmental degradation and the use of pesticides in the United States. The ripple effect was that pretty much over night the awareness of human-caused damage to the environment had entered public discourse. While at first the approach was inherently human-centred (anthropocentric), over time it became more and more interested in the value of the environment and its components itself. This biocentric approach *vis-à-vis* anthropocentric perceptions of environmental protection constitute a significant normative and ideological divergence in current environmental law regimes.

While Carson's book triggered the modern version of international environmental law, this is not to say that after its publication the first conservation regimes were adopted. Far from it. As we will see, already in the nineteenth and early twentieth century international cooperation existed for the protection of specific environments

[2]Churchill and Lowe (1999).

[3]Couzens (2013).

[4]Koivurova (2014).

[5]Carson (1962/1994).

or specific species. Seals played a major role in this regard, as Chap. 6 will demonstrate. Particularly in the colonies of the European states, the colonisers put in place legal regimes to protect the environment. For instance, the Convention Relative to the Preservation of Fauna and Flora in their Natural State[6] established national parks and strict natural reserves in colonial Africa. The point of this convention was not to protect the environment for the sake of the environment. Instead, it was to provide for as little disturbance as possible in order to retain its natural state so that colonisers had the possibility to 'escape' the noisy life in the city. This being said, the 1930s were a time when, surprisingly, also the first principles of international environmental law were codified. The principles of environmental law are as such not enforceable rules, but rather underlying norms that have found their way into the provisions of different legal regimes. The first ones that were formulated and generally accepted were the avoidance of transboundary pollution and the associated 'polluter pays' principle as well as the 'no-harm' principle.

These principles had their origins in the *Trail Smelter* case which revolved around a smelter in Canada that emitted sulphur dioxide, damaging crops and property in the United States. Since neither legal system provided for legal remedies, the US and Canadian governments relied on an arbitration tribunal to resolve this issue. Throughout years of legal proceedings, in 1941 the tribunal finally ruled that a state cannot use its own territory in a way that would cause harm to others. Ultimately, Canada was therefore ordered to pay compensation to the United States. The no-harm principle has become widely known as *Sic utere tuo ut alienum non laedas* (or abbreviated to *sic tuo utere*) which translates into "use your own property in such a manner as not to injure that of another." This principle would become relevant internationally some years later.

With the establishment of the United Nations (UN) in 1945, the path for further cooperation on all different kinds of issues was paved. Also global environmental cooperation was consequently made easier. At first, however, the prevention of war, economic growth and the promotion and advancement of human rights stood at the core of the UN. Consequently, the environment was not considered as crucially important other than a resource. The 1949 UN Conference on the Conservation and Utilization of Resources, for example, was convened for the purposes of exchanging information on resource exploitation and for increasing capacities.

By then, of course, the international community had realised that humankind is capable of causing wide-scale environmental destruction and indeed total annihilation of life on Earth—best exemplified by the short- and long-term effects of the nuclear bombs on Hiroshima and Nagasaki in 1945. Only 3 years after the end of the Second World War, the International Union for the Conservation of Nature (IUCN) was established, consisting of governments and civil society organisations. The IUCN at first examined the impacts of human activities on nature, especially as regards the role of pesticides. Over time it would develop into a core international

[6]Convention Relative to the Preservation of Fauna and Flora in their Natural State, 8 Nov 1933 (1 LNTS 172).

player and an authority responsible for motivating and ultimately drafting key international regimes for the conservation of the environment.

We must remember that up until the 1960s, issues of environmental protection were mostly in the hands of the industrialised nations, which largely corresponded to the world's colonial powers. For instance, the first global environmental protection regime, the 1954 International Convention for the Prevention of Pollution of the Sea by Oil,[7] was a regime signed and ratified by mostly western states. Only throughout the 1960s, when the decolonisation process enabled the shift of sovereign powers to former colonies, the international legal landscape changed, making so-called 'developing nations' active players in international law-shaping and -making. Particularly their contributions to the negotiations of the United Nations Convention on the Law of the Sea (UNCLOS)[8] between 1973 and 1982 is noteworthy. We have to bear in mind that at that time even far-reaching and rather progressive regimes were mostly negotiated by western governments.

This being said, when in 1972 the United Nations convened the Conference on Environment and Development in Stockholm (Stockholm Conference), the shift from western-dominated environmental law-making towards a truly global effort started to take form. It is at least here that also international environmental law saw the manifestation of legal principles. First and foremost, sovereignty or the sovereign rights over natural resources. Although it had become clear that international cooperation on environmental matters was necessary, this would not mean that countries would lose their sovereign rights to their own territories and resources. While sovereignty was manifested already in the Charter of the United Nations, in environmental law contexts this would mean that other states would not be able to force a government to pursue a specific environmental protection strategy. Of course, a broad body of literature on international law and policy-making that also tackles issues of coercion exists. However, as a fundamental principle, each state is entitled to its own social and legal system for environmental protection. Nevertheless, the *sic tuo utere* principle puts limits to this sovereignty. The Stockholm Conference confirmed this principle, which is also known as the principle of good neighbourliness.

Throughout the 1970s and 1980s the geopolitical situation in the world started to change once again. Glasnost and perestroika in the Soviet Union started to lift the 'Iron Curtain', ultimately resulting in the collapse of the Soviet regime in 1991. Only 1 year later the United Nations convened the United Nations Conference on Environment and Development (UNCED) in Rio de Janeiro (Rio Summit). This meeting is another a landmark event for the protection of the environment. Several agreements are a direct outcome or were profoundly motivated by the summit: the Convention on Biological Diversity[9]; and the United Nations Framework Convention on Climate

[7]International Convention for the Prevention of Pollution of the Sea by Oil, 12 May 1954.

[8]United Nations Convention on the Law of the Sea, 10 Dec 1982 (1833 UNTS 3).

[9]Convention on Biological Diversity, 5 Jun 1992 (1760 UNTS 79).

Change (UNFCCC)[10]; the United Nations Convention to Combat Desertification (UNCCD)[11]; and the so-called Straddling Fish Stocks Agreement.[12]

Apart from these conventions, the Rio Summit furthermore established the principle of sustainable development. Drawing from the report of the World Commission on Environment and Development (WCED), chaired by Gro Harlem Brundtland, entitled *Our Common Future* (oftentimes referred to as the Brundtland Report),[13] sustainable development is defined as 'development that meets the needs of the present without compromising the ability of future generations to meet their own needs.' In order to achieve this, the principle of sustainable development includes several sub-principles: sustainable use, which is as such not a new principle but for the first time integrated in an international legal framework; intergenerational equity; and integration of environment and development.

The principle of intergenerational equity sees one generation merely being entrusted with a certain wealth the same amount of which (if not more) it would pass on to the next generation. Very closely hand in hand, yet conceptually different, is the principle of intragenerational equity, which sees the same logic between different peoples of the same generation. The principle of sustainable use, a key principle in the contexts of marine mammals, sees that resources are used in a sustainable manner, avoiding overexploitation and resource depletion. No commonly accepted definition of this principle exists and it appears in different forms—for instance wise use, rational use, sound environmental management—in environmental law.

The principle of sustainable development is also referred to as the 'precautionary principle' or the 'precautionary approach.' In essence this means that when it comes to the exploitation of natural resources, a level of precaution is to be applied: if scientific evidence is uncertain, it is better to refrain from a certain activity rather than risking environmental damage. While in theory logical and comprehensible, in practice different countries attach different values to scientific uncertainty, reach different results or interpret scientific findings differently. Perhaps even more so, scientific results are seen through a political lens. This is also the case for marine mammals, and in particular whales and whaling (see Sect. 5.3.4). Consequently, a scientific reality or 'truth' which different actors interpret through different lenses is difficult to establish.[14] Instead, I would rather consider scientific findings part of different realit*ies* that generate different truth*s* for countries and peoples all over the world.

---

[10]United Nations Framework Convention on Climate Change, 9 May 1992 (1771 UNTS 107).

[11]United Nations Convention to Combat Desertification in Countries Experiencing Serious Drought and/or Desertification, Particularly in Africa, 17 June 1994 (1954 UNTS 3).

[12]United Nations Agreement for the Implementation of the Provisions of the United Nations Convention on the Law of the Sea of 10 December 1982 relating to the Conservation and Management of Straddling Fish Stocks and Highly Migratory Fish Stocks, 4 August 1995 (2167 UNTS 88).

[13]WCED (1987).

[14]Heazle (2006).

The principle of integration considers that the protection of the environment, and economic and social development cannot be separated. Organisations such as the World Trade Organization (WTO), for example, did not consider this interlinkage when they were established. Instead, the WTO, to stick with this example, merely focused on free trade and associated economic growth. Only over time has the WTO has started to consider environmental standards and principles and, as for example Timo Koivurova argues, should also be considered within contexts of international environmental law.[15] Indeed, the WTO is an international organisation which we will also get to know in the controversies surrounding the European Union's regime on trade in seal products (see Sect. 6.6).

The principle of common but differentiated responsibilities is a principle which is probably the oldest principle in international environmental law, even though it has not been formulated as such until the Rio Conference. The principle in contemporary discourse refers to the Global North-South axis, which aims to lift the burden of developing nations to make due for the environmental harm caused by industrial nations. We can find the same reasoning applied in other regimes, dating back more than one hundred years in the form of 'aboriginal/indigenous/native exemptions' to sealing regimes, as we will see in Chap. 6. The logic is essentially the same: aboriginal peoples are to be spared from limits to the seal hunt since their small-scale subsistence hunt does not significantly contribute to the population decline of seals. This is contrary to the large-scale hunts of commercial enterprises which decimate the population. In other words, aboriginal peoples should carry a different degree of responsibility towards seal conservation than commercial enterprises (Table 1.1).

With this background knowledge on international (environmental) law, I now proceed to the basics on marine mammals in the next chapter and the history of interaction between humans and marine mammals in Chap. 3.

Table 1.1 Core principles of international environmental law

Avoidance of transboundary pollution/No-harm principle	Use your territory in a way that it does not cause damage to others
Polluter pays	The originator of environmental harm has to bear the financial and legal burden
Sustainable development	Use the environment in a way that it is abundant in the same, if not improved, way for generations to come
Common but differentiated responsibilities	Countries less responsible for long-term environmental damage should bear a lesser burden for its remedy

[15] Koivurova (2014), pp. 144–146.

References

Carson R (1962/1994) Silent spring. Houghton Mifflin Company, Boston
Churchill RR, Lowe AV (1999) The law of the sea, 3rd edn. Manchester University Press, Manchester
Couzens E (2013) Whales and elephants in international conservation law and politics. A comparative perspective. Routledge, Abingdon
Heazle M (2006) Scientific uncertainty and the politics of whaling. University of Washington Press, Seattle
Koivurova T (2014) Introduction to international environmental law. Routledge, Abingdon
Sellheim N (2018) The seal hunt. Cultures, economies and legal regimes. Brill Nijhoff, Leiden
WCED (1987) Our common future. United Nations, Geneva

Further Reading

D'Amato A, Chopra SK (1991) Whales: their emerging right to life. Am J Int Law 85:21–62
Dorsey K (2013) Whales & Nations. Environmental diplomacy on the high seas. University of Washington Press, Seattle
Fitzmaurice M (2016) Whaling and international law. Cambridge University Press, Cambridge
Fitzmaurice M, Tamada D (eds) (2016) Whaling in the Antarctic. Significance and implications of the ICJ judgment. Brill Nijhoff, Leiden
Glenn HP (2014) Legal traditions of the world, 5th edn. Oxford University Press, Oxford
Jefferies CSG (2016) Marine mammal conservation and the law of the sea. Oxford University Press, Oxford
Koremenos B (2016) The continent of international law. Explaning agreement design. Cambridge University Press, Cambridge

Marine Mammals: Some Basics

2

The focus of this book rests on a group of mammals, which are commonly known as marine mammals since their habitats can be found in the marine environment. Under this category fall five different orders: whales (*Cetacea*); seals (*Pinnipedia*); the polar bear (*Ursus maritimus*); sea cows (*Sirenia*); and otters (*Lutrinae*).[1] Especially the latter has played a fundamental role in the colonisation process of the American Arctic since otter furs were a much sought-after trade good which led to Russian traders, settlers and colonisers setting foot in Alaska. Since, amongst other things, the trade in otter furs was for several reasons not sustainable and consequently not sufficiently profitable, Russia lost interest in Alaska and sold it to the United States in 1867. Yet, both otters and sea cows will only play a minor role in this book since they no longer play a prominent role in international law.

This chapter focuses on three orders of marine mammals—*Cetacea*, *Pinnipedia*, *Ursidae*—and presents some key species that have played a role in both their exploitation and conservation. I have chosen to include a brief introduction on these species since oftentimes marine mammals, first and foremost 'the whale' and also the polar bear, are misrepresented or mischaracterised in the public discourse. This is to say that 'the' whale does not exist, but must be considered on a species to species or even population to population (stock to stock) basis. 'The' whale as an overarching species is therefore not necessarily intelligent, endangered or sings songs and we will delve into this characterisation in the next chapter. Also, the claim that the polar bear is endangered and a victim to climate change is simplified at best. As we will see, polar bears should rather be considered in plural and their population trends are far from being clear.

Since there are around at least 100 different marine mammal species (probably more, depending on the classification), it is not possible to describe each single one of them. Therefore merely those whale and seal species that have been most prominent in the public discourse, as well as the polar bear, will be presented here.

[1]Evans and Raga (2001), pp. 1–7.

N. Sellheim, *International Marine Mammal Law*,
https://doi.org/10.1007/978-3-030-35268-4_2

2.1 Whales

Whales (*Cetacea*) are probably the most well-known group of marine mammal species. Depending on taxonomic classification, between 80 and 90 different species exist. By and large it is assumed that at least 84 different cetacean species exist, although new classification methods allow for more detailed analyses and determination of species. Whales are being subdivided into two subgroups: baleen whales (*Mysticeti*) and toothed whales (*Odontoceti*). Baleen whales numerically constitute the smaller group of whales although this group contains the largest known mammals on Earth. The blue whale (*Balaenoptera musculus*), for instance, with 33 m body length holds the record of the largest living animal.

Baleen whales themselves are subdivided into three sub-families: rorquals (*Balaenopteridae*), right whales (*Balaenidae*), and gray whales (*Eschrichtiidae*). Additionally, another family might exist, *Cetotheriopsidae*, which the pygmy right whale (*Caperea marginata*) might be a member of. Yet, this is subject to further inquiry since genetic studies have not succeeded in fully agreeing on the classification of this whale species.

Toothed whales comprise at least 71 species, which are subdivided into several sub-groups: dolphins (*Delphinidae*), beaked whales (*Ziphoidae*), sperm whales (*Physeteroidae*) and *Monodontidae*, which comprises the narwhal (*Monodon monoceros*) and the beluga whale (*Delphinapterus leucas*). The genetic affiliation of river dolphins has not been conclusively been resolved. Toothed whales are rather small cetacean species. The Hector dolphin (*Cephalorhynchus hectori*), for example, reaches a maximum length of 1.5 m. The exception are the sperm whale whose males can reach up to 18 m in length, as well as larger species of beaked whales (*Berardius* and *Hyperoodon spp.*).[2]

In public discourse, 'the whale' is often ascribed human-like, anthropogenic characteristics. 'The whale' is thus a social animal of high intelligence, has developed characteristics that some classify as cultures, and it sings songs for communication. Moreover, 'the whale' is critically endangered because of commercial exploitation in the past and due to climate change. These narratives are widely reflected in a multitude of different socio-cultural outlets, particularly in the Western world, and have significantly influenced the discourse on whales and whaling throughout the 1960s and 1970s.[3] The dissemination of whale songs in modern pop culture is a reflection of this presentation of whales. The Norwegian anthropologist Arne Kalland has coined the term 'superwhale' in this context, which negates the diversity of whale species and which only exists in human consciousness.[4] Along with this narrative goes the view that from an ecological and moral perspective, whales should not be killed for human consumption. In public discourse the consumption of whale meat is, at least in the Western world, not acceptable.

[2]Ballance (2008), pp. 196–201.

[3]Epstein (2008).

[4]Kalland (2009).

Irrespective of the cultural, social and intelligence-related skills and the associated moral implications, the following pages reflect the ecological state of some whale populations. However, since data collection and analysis are extremely difficult due to whales spending most of their time under water and being often migratory, the information presented here is not absolute. Still it serves as a backdrop for the complexity of the interaction between marine mammal ecology and international law.

2.1.1 Blue Whale (*Balaenoptera musculus*)

The blue whale (*Balaenoptera musculus*) belongs to the family of baleen whales and is the largest living species. The habitat of the blue whale extends over all oceans. For this species, females are being used to determine the maximum body size since they display a significant difference in size compared to males (sexual dimorphism). The largest blue whales were thus females, caught in the waters surrounding the South Shetland Islands and South Georgia, reaching sizes between 31.7 and 32.6 m. In 1947, a female of 190 tons was caught at South Georgia. The normal weight of blue whales ranges between 50 and 150 tons.

Although reference is commonly made to 'the' blue whale, the species is subdivided into at least three sub-species: *B. musculus intermedia,* whose habitat is in the Antarctic; *B. musculus brevicauda* with a habitat in Antarctic waters, in the Indian and Pacific Oceans, commonly referred to as the pygmy blue whale; and *B. musculus musculus* in the northern hemisphere. Rather dense populations can be found in the North Atlantic and along the west coasts of North and South America.

We must note, however that the term 'dense population' is relative. After all, since the beginning of the nineteenth century the blue whale has been excessively hunted due to its large size and associated commercial value. This resulted in a drastic reduction of the population. For instance, in the hunting season of 1930–1931 almost 30,000 animals were killed. It is estimated that in the first half of the twentieth century 325,000–360,000 blue whales were killed. This led to the blue whale populations being close to extinction. Only in 1963 the International Whaling Commission placed the blue whale under strict protection in waters south of 40° south latitude and 0–80° east longitude.

Due to the importance of the blue whale in international trade and in respect to the dwindling populations, in 1945 the so-called blue whale unit (BWU) was introduced. The BWU served as a reference for all hunted whales. One blue whale corresponded to two right whales, two and a half humpback whales or six sei whales. Although the BWU was used as an attempt to protect the blue whale it had the opposite effect: whalers targeted larger animals to reach their assigned BWU limits more quickly. As a consequence, the populations of blue, grey, right and humpback whales were massively reduced. Only in 1972 the BWU was abandoned and replaced with species-based quotas.[5]

The current population of the blue whale is labelled as 'endangered' by the IUCN although the population trend is increasing. It is estimated that there are now around 10,000–25,000 adults worldwide, corresponding to around 3–11% of the original population size.[6] Since the populations of blue whales are small, all whaling nations have banned the hunt for blue whales for commercial or other reasons.[7]

2.1.2 Fin Whale (*Balaenoptera physalus*)

The fin whale (*Balaenoptera physalus*) is a close relative of the blue whale and it is estimated that there is at least one hybrid amongst 500–1000 fin whales. Only recently the high degree of hybridity between these two species has been made public when Icelandic whalers seemingly landed a blue whale which is strictly protected under Icelandic law. Upon closer genetic inspection, this whale, however, proved to be a hybrid between blue and fin whale.[8]

As in the case of the blue whale, sexual dimorphism in fin whales also favours females since they are on average 5–10% longer than males. Fin whales in the southern hemisphere reach as much as 25–26 m while their northern relatives are around 21–22.5 m long. The mass of the fin whale ranges between 40 and 50 tons in the northern hemisphere and 60–80 tons in the south. The habitat of the fin whale extends into all oceans, except the polar regions. Up to the point of writing, no fin whales have been spotted in the Black Sea and are rare in the eastern Mediterranean, in the Baltic Sea, in the Persian Gulf, in the Red Sea and close to the equator. The highest population density can be found in temperate and cold water, although we must remember that the fin whale is a highly migratory species. It is surmised that fin whales form single independent groups for reproduction. Since the fin whale is a fast swimmer with a large body size, apart from the killer whale/orca (*Orcinus orca*) the fin whale does not have natural predators.[9]

Throughout the nineteenth and twentieth century the fin whale was one of the most overhunted whale species. Within a time period of approximately 30 years in which the fin whale was hunted in the North Atlantic, its population had declined to such a degree that already in 1904 whaling stations which were specialised on fin whales had to close down. As a consequence, the hunt for fin whales moved into other regions of the North Atlantic, then into the Antarctic, the Pacific and finally into the Indian Ocean. For the years 1935–1970 it is estimated that 30,000 fin whales were killed annually. Only with the coming into force of the whaling moratorium in

[5]Gillespie (2005), p. 4.

[6]Cooke (2018a).

[7]Sears and Perrin (2008), pp. 120–124.

[8]McGrath (2018).

[9]Aguilar (2008), pp. 433–437.

the year 1985/86 (see Sect. 5.3.4) fin whale populations have started to recover. Since 2006 Iceland has actively been hunting fin whales with a quota of 160–190 individuals per year. Up until Japan's withdrawal from the International Whaling Commission in December 2018 it had been hunting 3–10 fin whales per year in the Antarctic for scientific purposes.

The IUCN lists the fin whale as 'vulnerable.' It is not clear, however, which fin whale population is subject to which specific threat. For the North Atlantic a population size of more than 50,000 animals is estimated. The global fin whale population is assumed to be around 100,000 animals and increasing.[10] However, these numbers must be used with care since evaluation and assessment methods are imprecise.

2.1.3 Bowhead Whale (*Balaena mysticetus*)

The bowhead whale (*Balaena mysticetus*) can only be found on the northern hemisphere close to the Arctic sea ice. Female bowhead whales are larger than males and can reach a length of 14–17 m and a weight of 75–100 tons. The head of this whale species constitutes around one third of the overall size while the baleen reaches a length of up to 4 m. The combination of skin and blubber is the thickest amongst the whales: it is around 2.5 cm thick and merges with a layer of blubber of 5.5–28.5 cm thickness. The characteristic bow-like shape of its mouth is the patron of its name.

The global population of bowhead whales is subdivided into four stocks: the Western Arctic or Bering-Chukchi-Beaufort stock around Alaska; the Okhotsk stock in Eastern Russia; a stock in northeastern Canada in the Davis Strait and Hudson Bay, and around Svalbard in the North Atlantic.[11]

Since the sixteenth century, bowhead whales have been hunted intensively, resulting in a drastic reduction of the overall population. However, since the end of the commercial hunts in 1915 all stocks have recovered. Indigenous peoples in Alaska and Chukotka are legally entitled to a small number of bowheads under the auspices of the International Whaling Commission. The overall number of hunted bowhead whales does not exceed 50 individuals.

The IUCN lists the bowhead whale as 'least concern' with an increasing population trend. Although the overall population status is not finally resolved, it is assumed that it at least comprises 100,000 animals.[12]

---

[10]Cooke (2018b).

[11]Jungh and Shelden (2008), pp. 131–133.

[12]Cooke and Reeves (2018).

2.1.4 Right Whale (*Eubalaena spp.*)

The order *Eubalaena* comprises of three types of right whale: the North Atlantic right whale (*Eubalaena glacialis*), the North Pacific right whale (*Eubalaena japonica*), and the southern right whale (*Eubalaena australis*). The species are close relatives of the bowhead whale and are also very similar in look. Right whales, however, display a high degree of callosity on their heads, the jawline, the chin and around the blowhole. These are caused by parasitic crustaceans, commonly referred to as 'whale lice'.

As is in the case of bowheads, the head of the right whale comprises around one third of the body. The size of right whales ranges between 13 and 18 m whereas northern right whales are generally larger than their southern relatives. Also right whales display a bow-like mouth shape, but are easily distinguishable from bowhead whales due to the characteristic callus. Different shark species as well as killer whales predate on right whales.

Traditionally, the populations in the North Atlantic have been subdivided into a western and eastern population although there is no genetic difference between them. Sightings of individual whales in the eastern Atlantic which could be identified as being part of the western population underline this. Historically, stocks of right whales could be found in the waters from Florida to Labrador, in southern Greenland, around Iceland and northern Norway as well as in western Africa. Nowadays, the western Atlantic stock ranges from Florida to Nova Scotia and rarely beyond and constitutes the main Atlantic stock since hardly any sightings have occurred in the eastern Atlantic in recent times.

The habitat of the North Pacific right whale extends traditionally from Japan, Northern Mexico, the Sea of Okhotsk, the Bering Sea and the Gulf of Alaska. Accurate data on the population status of the North Pacific right whale is difficult to obtain since the stock has decreased significantly.

The Southern right whale is subdivided into several stocks although a final determination of these stocks has yet to occur. The current state of research indicates that separate stocks exist in Argentina and Brazil, South Africa, east Africa and Mozambique, western Australia, south-eastern Australia, New Zealand and Chile. While only the stocks in Argentina, South Africa and Australia are well documented, concrete numbers on the overall population are not available.

The right whale was the first whale which was hunted by Basque whalers in the fifteenth century, due to the ease with which the whale could be hunted. Its slow motion and the fact that it floated after it had been killed earned it the name 'right' whale as it was indeed the 'right' whale to hunt. Although it is not clear whether Basque whaling beyond the Bay of Biscay has significantly impacted the right whale population, genetic studies of the approximately 40,000 whales that the Basques landed in the seventeenth century were predominantly North Atlantic right whales. When commercial coastal whaling developed in North America throughout the eighteenth century, right whale populations were reduced drastically. This was also caused by the expanding US whaling operations ('Yankee whaling') in the North and South Pacific as well as in the South Atlantic. Approximately 38,000 right whales were killed in the South Atlantic, 39,000 in the South Pacific, 15,000 in the North Pacific, 1300 in the Indian Ocean and several hundred in the northeastern Atlantic. When the first whaling convention was concluded in 1931, all right whale hunts were suspended.[13]

Despite the IUCN listing the South Pacific right whale as 'least concern', both its population trend and overall population status are unknown.[14] Both the North Atlantic and North Pacific right whale are considered 'endangered' although also here concrete numbers are absent.

2.1.5 Humpback Whale (*Megaptera novaeangliae*)

The humpback whale (*Megaptera novaeangliae*) probably belongs to one of the most famous whale species in the world and is much sought-after by whale tourists due to its typical leaps out of the water, also close to ships and boats. Since the humpback whale stretches its tail fluke into the air when diving, scientists have been able to identify individual whales due to their characteristic markings on their tail flukes.

The humpback whale has very large pectoral fins, proportionally the largest amongst the whales, amounting to almost one third of the body length. The scientific term *Megaptera* derives from the Greek term *mega* (large) + *pteron* (wing). The area around the mouth is marked by a large number of nobbles. The dorsal fin is shaped in a humpy (German, *Buckel*) manner, which has provided the whale with both the English as well as the German name (*Buckelwal*).

Female humpback whales are typically 1–1.5 m longer than males. The maximum length of these animals ranges between 16 and 17 m although most humpbacks are 14–15 m long. In the summer, humpback whales can mostly be found in coastal and shelf waters while in the winter, in waters around reefs and islands. Humpback whales are highly migratory and it is a humpback which is believed to hold the longest whale migration: 5000 km. Interestingly, the purpose and aim of humpback migration still remains unclear since not all populations migrate along similar patterns. The stock in the Arabian Sea, for instance, does not migrate at all. It remains clear, however, that humpbacks return to specific breeding grounds each year.

Since humpbacks can often be found close to shore they constituted one of the main targets of commercial whaling operations over several centuries. Given their global distribution, they also served as a primary food source during colonial endeavours. In the southern hemisphere, around 238,000 humpbacks were killed in the twentieth century, 48,000 of which were illegally taken by the Soviet Union. Although humpback whaling was amongst the first types to be banned, in the whaling season 1959/60 and 1960/61 the Soviet Union killed around 25,000

---

[13]Kenney (2008), pp. 962–969.

[14]Cooke and Zerbini (2018).

humpbacks, all in all amounting to around 90% of the overall population back then.[15]

The current global population of humpback whales is estimated to be at least 84,000 animals with an increasing population trend. The IUCN therefore categorises the humpback as 'least concern.'[16]

2.1.6 Gray Whale (*Eschrichtius robustus*)

The gray whale (*Eschrichtius robustus*) is the only whale species in the genus *Eschrichtius*. Before the last Ice Age, the North Atlantic and North Pacific populations underwent some mixing via the Northwest Passage. When the waters of the passage started to freeze over and ultimately became perennially frozen, separated populations with unique genetic features started to develop. In the late seventeenth or early eighteenth century, the gray whale could be found in the north Atlantic and North Pacific Ocean until the Atlantic population disappeared. While Icelandic, Basque and Yankee whaling certainly contributed to the decline of the Atlantic gray whale population, it is unclear whether whaling constituted the only reason for the disappearance of the population. Contemporary gray whale populations are limited to the Pacific Ocean where two extant populations can be found. The eastern North Pacific and adjacent Arctic Ocean waters is the habitat of the American/California/Chukchi stock. The Asian/Korean/Okhotsk stock can be found in the North Pacific off the Asian coasts.

Gray whales have always been on the top of the list of commercial whalers since they move rather slowly and are therefore relatively easily to be hunted. Consequently, gray whale stocks declined significantly over the nineteenth and early twentieth century. When the International Whaling Commission was founded in 1946, gray whales were amongst the first species to be protected and commercial whaling on gray whales was banned. This did not account for aboriginal subsistence whaling. Currently, the Chukchi of the Chukotkan Peninsula in the Russian Federation and the Makah in Washington State in the United States conduct aboriginal subsistence whaling, also targeting gray whales. The quotas allocated for gray whale hunts are also shared amongst the Chukchi and Makah. Although gray whale hunts had been banned since 1946, this did not mean that gray whales had been continuously protected. In the 1960s and 1970s, the Soviet Union did not adhere to the prohibition of gray whale hunting and killed nearly every gray whale upon sight. Also the deliberate and accidental by-catch of gray whales in fishing nets by Japanese fishers puts tremendous pressure on the population.

The gray whale displays several unique characteristics. On the one hand it has been termed the 'devil fish' by whalers due to its ferocity when being hunted. On the other, it is much sought-after by the whale watching industry due to its friendliness.

[15]Clapham (2008), pp. 582–585.

[16]Cooke (2018c).

Another (passive) characteristic has been found in numerous gray whales since the 1960s: foul odour, which makes the meat inedible for both humans and animals. These so-called 'stinky whales' occur only on the Chukotkan Peninsula and have been reported by Chukchi whalers. The cause of this phenomenon is unknown but it is surmised that whale metabolism has changed due to limitations in the food supply and availability, different or new foods, biotoxins or disease. Notably, also some seals, walrus, fish and murre eggs display this phenomenon. Between 1999 and 2007 an additional phenomenon referred to as 'skinny whales' could be found. These emaciated whales were of extremely poor condition but the causes also remain unknown. It is possible that habitat degradation, changes in the availability of food, stress or disease may have caused the 'skinny whale' phenomenon.[17]

The IUCN lists the gray whale as 'least concern' since its population status has recovered over the last three generations, despite some fluctuation. This, however, only accounts for the eastern North Pacific Ocean population. The western North Pacific population is possibly extinct.[18] Throughout the spring and summer of 2019, an unusually large number of gray whale carcasses could be found on the coasts of North America. Whether this was a one-time mortality event is under investigation at the time of writing. Yet, since the dead whales were emaciated, scientists believed food shortage and possibly the effects of warming waters have contributed to this event.

2.1.7 Sperm Whale (*Physeter macrocephalus*)

The sperm whale is the only species in the genus *Physeter*. Although in the past there was some disagreement over the genetic affiliation of sperm whales or whether there are in fact several species of *Physeter*, the contemporary view is that there is merely one species, *Physeter macrocephalus*.

The sperm whale displays unique characteristics and is a 'poster species' for what constitutes the 'superwhale': it is very large and holds a very large brain; it lives very long lives with no predators other than orcas and humans. This has allowed the sperm whale to be able to develop deep social bonds with other sperm whales, which have made the effects of hunting go beyond the actual animals killed; it dives extremely deep and thus spends most of its time under water. The sperm whale is the whale species which in many ways resembles the African elephant (*Loxodonta africana*) physically and socially. One of the main features of the sperm whale is the spermaceti organ, a complex organ in the massive nasal complex of the whale. This organ which comprises spongy tissue may have contributed to being able to gain access to a wide array of food and may be the cetacean equivalent to the trunk of the elephant. This, however, is not the only parallel and it is particularly the complex

[17]Jones and Swartz (2008), pp. 503–511.

[18]Cooke (2018d).

social structures, mating behaviour and other characteristics which make the elephant and the sperm whale appear to be relatives.

Sexual dimorphism is prominent in sperm whales. While males reach sizes of around 16 m and 45 tons, females grow up to 11 m with a weight of around 15 tons. The sperm whale has a global distribution and both males and females can be found in productive, temperate waters in all oceans. The older the male sperm whales get, the higher their average latitude. This means that old sperm whales can also be found along the edges of the pack ice in both the Arctic and the Antarctic. Whilst they do return to more temperate waters, the timing by which this occurs remains unknown.

Sperm whales were a primary target of the commercial whaling industry the nineteenth century since their high fat content and the oily spermaceti organ were valuable drivers of the industrial revolution. Particularly Yankee whalers chased sperm whales all over the world and Herman Melville's epos *Moby Dick* describes the life on board an American whaler in pursuit of the white (sperm) whale.

With the discovery of hydrocarbon fuels, the oil of the sperm whale lost its commercial importance and towards the first half of the twentieth century sperm whale hunts had declined. After the Second World War, however, both the oil and the meat of the sperm whale gained significance again. The IWC permitted sperm whaling, yet under specific conditions, until the whaling moratorium was adopted in 1982. Merely Japanese scientific whaling in the Antarctic as well as small-scale whaling off the coasts of Indonesia constituted the only ongoing sperm whale hunts until Japan's withdrawal from the IWC, when Antarctic whaling was halted. Nowadays, only some Indonesian islanders take a small number of sperm whales per year. The sperm whale is the only toothed whales that falls under the purview of the whaling moratorium.[19]

Although the sperm whale was hunted massively, it appears that its population has recovered to a degree that it can no longer be considered endangered. The IUCN lists the sperm whale as 'vulnerable' since both the population trend and the state of the overall population are not known.[20]

2.1.8 Common Bottlenose Dolphins (*Tursiops truncatus*)

When 'the' dolphin is referred to, it usually means the common bottlenose dolphin (*Tursiops truncatus*), which is arguably one of the most commonly known marine mammals in the world. The bottlenose dolphin has a global abundance and can be found in all tropical and temperate waters. Its northernmost habitat stretches towards the Lofoten Islands in northern Norway, the water at Nova Scotia, the Kuril Islands and the Sea of Okhotsk. In the south, bottlenose dolphins can be found in Terra del Fuego, Australia, New Zealand and South Africa.

[19]Whitehead (2008), pp. 1091–1097.

[20]Taylor et al. (2008).

The size of this species ranges from 2.5 to 3.5 m, depending on geographic location. Since there are physiological differences between bottlenose dolphin populations, this has caused disagreements over the genetic composition of the species. It was assumed that there are around 20 different bottlenose species, however contemporary scientific consensus is that there is one genus, although a re-evaluation of this state of affairs is currently ongoing. The fact that bottlenose dolphins and other toothed whales appear to interbreed makes the genetic differentiation increasingly difficult.

Bottlenose dolphins have displayed different migratory behaviours. While some populations migrate, others are periodic resident species, or combine both migration and local residency. They can often be found in coastal waters, in bays and estuaries, and have also been sighted in rivers. Usually, bottlenose dolphins form groups of 2–15 individuals although very large groups of up to 1000 animals have been observed. Amongst the groups, strong familial ties, sexual relationships and other forms of individual affiliation exist. Like other dolphin groups, the bottlenose dolphin communicates and navigates through clicking (echolocation), whistling (individual sounds to communicate identity and location) and burst-pulse sounds ("squawks" during social interactions).

Bottlenose dolphins have displayed a high degree of intelligence and make frequent use of human infrastructure. For example, dolphins have been observed driving schools of fish towards fishers in order to be able to hunt them more easily. Bottlenose dolphins have also been trained for military purposes. They can frequently be seen following fishing trawlers, collecting discarded fish, or stealing fish from fishing gear. Since the nineteenth century, but particularly during the twentieth century, bottlenose dolphins have been part and parcel of aquariums and zoos worldwide and dolphin shows are a common attraction (see Fig. 2.1).[21]

One of the main contemporary sources for bottlenose and other dolphins is the Japanese town of Taiji where the drive fishery is conducted for food and for capturing dolphins for the international markets. Since 2015 the World Association of Zoos and Aquariums (WAZA) has ordered its members to stop purchasing dolphins from Taiji, which has resulted in some Japanese aquariums leaving the Japanese branch of the organisation. While the capturing of dolphins was one reason, Taiji's large-scale and well-documented drive fishery is another. The Oscar-winning documentary-like film *The Cove* by Louie Psihoyos from 2008 is an important source for the influence on the public perception of the drive fishery. Although several cinematic attempts have been made to counteract the claims made in *The Cove* and even though the Japanese government has tried to justify the drive fishery in the International Whaling Commission, international opposition that has translated into pressure on the Japanese government is significant.[22]

However, Japan is not the only country hunting or having hunted bottlenose dolphins. In the Black Sea, dolphins had been hunted extensively but the Black Sea

[21]Wells and Scott (2008), pp. 249–255.

[22]Sellheim (2017).

Fig. 2.1 Bottlenose dolphins in the Taiji Whale Museum © Nikolas Sellheim

littoral states have all now outlawed the dolphin hunt. In the Faroe Islands, dolphins are also taken as part of the annual pilot whale drive fishery (see next section). Although the overall population trend of the common bottlenose dolphin is not known, it is listed as 'least concern' by the IUCN. Primary threats to the species derive from diseases, pollution, ship strikes and underwater noise.[23]

[23]Wells (2019).

2.1.9 Pilot Whales (*Globicephala spp.*)

Pilot whales comprise two separate species, the long-finned pilot whale (*Globicephala melas*) and the short-finned pilot whale (*Globicephala macrorhynchus*). Male pilot whales can reach lengths up to 6 m while females are significantly smaller. Long-finned and short-finned pilot whales are difficult to distinguish by the way they look, but close examination allows a differentiation based on flipper length, skull shape or number of teeth.

The habitat of long-finned pilot whales are the colder temperate waters of the North Atlantic and the Southern Ocean. The two subgroups are genetically different and do not interact. Short-finned pilot whales can be found in tropical and subtropical waters and in the waters of the North Pacific. The short-finned pilot whales of northern and southern Japan, which are geographically separated, show differences in morphology, but the state of research assigns them both the same taxonomic name, *G. macrorhynus*.

Pilot whales have been extensively hunted all over the world. While in the past, drive fisheries have existed in Newfoundland, Norway, the Orkney Islands, Hebrides, and Iceland, contemporary drives are only conducted in the Faroe Islands and Japan. Although these hunts may take up to several hundred whales per year, there is no indication of population decline. What has become clear, however, is that pilot whales consumed by humans hold high levels of mercury. While in Japan these levels are not considered dangerous for humans, in the Faroe Islands the Faroese are advised to merely consume a limited amount of pilot whale per year to avoid negative effects of mercury. One of the other direct major threats to pilot whales is by-catch in different types of fisheries. It is unclear how large the numbers of pilot whales taken as by-catch are since only few countries are required to note pilot whale mortalities in fisheries.[24]

Both short- and long-finned pilot whales are listed as 'least concern' although their population trend is unknown.

2.2 Seals

Seals or pinnipeds (*Pinnipedia*) are a group of marine mammals comprising around 33 different species. Pinnipeds are subdivided into three subgroups: earless or true seals (*Phocidae*), eared or fur seals and sea lions (*Otariidae*), and the walrus (*Odobenidae*). Earless seals with 18 distinct species constitute the larges group of seals, followed by sea lions with 14 species. The walrus is the only member of the genus *Odobenidae*. Seals constitute around one third of all marine mammal species in the world.

Contrary to cetaceans, pinniped species have adapted to the marine, as well as the terrestrial, environment and their habitat can be found in both. This means that they

[24] Olson (2008), pp. 247–252.

spend their lives in both habitats and give birth to their young either on land or on sea ice. Also their physiology displays both aquatic and terrestrial features, reflected in thermoregulation, lactation, breathing and locomotion. On the other hand, seals can spend extended time under water, also in cold regions, as we will see below. Their physiology allows them to store large amounts of oxygen, to thermoregulate, to slow down their heart beats and to control spleen and cardiac functions while a thick layer of blubber and fur provide protection against the cold and against overheating on land. Their sensory organs are adapted to both life above and under water.

Earless seals are subdivided into two to four subgroup of seals, depending on the classification. We will not delve into this issue further. We find a distinction between *Phocinae* and *Monachinae*, also commonly differentiated as northern or southern phocids. Northern phocids comprise ten subspecies and can only be found in the northern hemisphere. Southern phocids comprise nine species with their main habitats in the southern hemisphere, three of which, however, can also be found in the northern hemisphere.

Eared seals are subdivided into fur seals and sea lions. Sea lions can be found in both the northern and southern hemisphere. Many fur seals have their habitat in the southern hemisphere. Apart from a small population off the coast of California—a prominent motive for tourists in San Francisco—the habitat of the northern fur seal (Callhorhinus ursinus) can be found in the North Pacific Ocean. We will get to know this species below. A small number of Guadalupe fur seals (*Arctocephalus townsendi*) can also be found in the northern hemisphere while the genus *Arctocephalus*—southern fur seal—is a southern hemispheric inhabitant. Southern fur seals can thus be found in South America, Australia, New Zealand and the (sub-)Antarctic.

The modern walrus (*Odobenus rosmarus*) is split into two distinct stocks, the Atlantic walrus (*Odobenus rosmarus rosmarus*) and the Pacific walrus (*Odobenus rosmarus divergens*). The habitat of the Atlantic walrus are the ice-covered, shallow waters of the Arctic in Canada, Greenland, Svalbard (Norway) and in the Laptev Sea. The Pacific walrus can be found in the Bering Sea and in the Pacific waters between Chukotka and Alaska.[25]

2.2.1 Harp Seal (*Pagophilus groendlandicus*)

The harp seal (*Pagophilus groenlandicus*) is probably the best-known pinniped species. Its prominence in particularly western discourse derives from the large-scale hunts in eastern Canada it has been subject to for centuries. Particularly in the second half of the twentieth century, harp seal pups were used as an icon against the seal hunt: white fluffy fur with large black eyes are indeed the perfect tool to display innocence (Fig. 2.2). In hunting and management parlance, harp seal pups are also referred to as ‘whitecoats’.

[25]Berta (2008), pp. 861–868.

Fig. 2.2 Harp seal pup (whitecoat) © IFAW

The harp seal shows a large black pattern on its back when adult, which resembles a harp. Harp seals can only be found in the northern Atlantic and Arctic Ocean. The overall population is subdivided into three distinct stocks: the northeastern stock in the Barents and White Seas; the 'Western' population in the Greenland Sea and around Jan Mayen; and the stock living and breeding in the northwest Atlantic along the east coast of Canada and in the Gulf of St Lawrence.

Harp seals migrate between Arctic waters in the summer and more temperate waters in the winter. In the spring season, they whelp their pups on the ice. This particular season has triggered the large-scale hunts for harp seals in Canada and around Jan Mayen. Whitecoats are born with a weight of around 11 kg. Over a period of just 12 days they are being nursed on extremely fat-rich milk, making them gain around 2.2 kg per day. Once the weaning period is over, the mothers abandon the young who then molt their fur and turn silvery in colour, referred to as 'beaters'. This stage lasts for around 12–14 months until the seal molts its fur again, changing it to a less-silvery fur colour. The seal is now referred to as 'bedlamer' (stemming from the French term *bête de la mer*, beast of the sea), lasting for several years (5.3 years on average) until the seal reaches sexual maturity. Upon the onset of sexual maturity, it once again molts its pelt into its adult stage, which it retains until its death.

The harp seal has been the target of human hunts for generations. Whilst it constitutes a primary food source for polar bears, orcas and sharks as well, it is the human which has been the species' main predator. During the eighteenth and nineteenth century, harp seals have been commercially hunted for their blubber and meat. In Newfoundland, during the nineteenth century, around one third of the economy was built on seals. At the highest peak in the middle of the nineteenth century, several hundred thousand harp seals were killed per year when seal oil was discovered to be a cheap alternative to whale oil as a driver of the industrial revolution. When hydrocarbons were discovered, the importance of seal oil diminished. Then, after the Second World War, whitecoat furs became an important commodity for the fashion industry. With the decline of international markets caused by a shift in public opinion concerning whitecoats and the associated first bans on whitecoat furs, the industry shifted towards beater furs. While the markets are considerably lower than 40 years ago, large numbers of harp seals are still landed for both the international and regional markets.[26]

The northwest Atlantic stock is estimated to have increased from 1.8 million animals in the 1970s to more than 7.5 million despite the still ongoing large-scale hunts. The stock around Greenland and Jan Mayen has increased from 296,000 in 1994 to 627,000 in 2013. The stock in the Barents and White Sea has dropped from 1.8 million in 2000 to 1.4 million in 2010 despite very limited hunts. The IUCN lists the harp seal as 'least concern' with an increasing population trend, but high fluctuations or pup production and mortality are a prevalent feature of this species. With the melting of the Arctic sea ice, harp seals lose important whelping grounds. Climate change poses therefore the greatest long-term risk for harp seals.[27]

2.2.2 Northern Fur Seal (*Callorhinus ursinus*)

The northern fur seal (*Callorhinus ursinus*) is one of the oldest genus of the family of *Otariidae* and its origins date back 2–5 million years. Northern fur seals display significant sexual dimorphism: while males range between 200 and 250 kg and have a black to reddish fur colour, females reach merely 45 kg and are grey-brown with a light underbelly. It is primarily a sub-Arctic seal species which can only be found in the Pacific Ocean. Around 74% of this species breed at the Pribilof Islands (St Paul and St George) in Alaska and 17% at the Commander Islands in Russia. The species can also be found at the Russian Kuril and Robben Islands as well as at the Channel Islands in California.

While the species spends considerable time on land to breed and to feed, it is also a migratory species. Despite interaction with humans dating back several hundred years and the fact this marine mammal species has been studied longer than any other, migration patterns and distribution, as well as the diet of particularly fur seal pups are not fully understood. This being said, it is clear that breeding preferences have largely remained the same over the last several hundred years, meaning that fur seals consistently return to the same areas on land in order to breed. These so-called 'rookeries' are well known by humans and have constituted the main sites for the large-scale seal hunts in the Bering Sea. Thirty-two of these rookeries are known to exist. Upon discovery of the species in 1742 by George Wilhelm Steller, exploitation began almost immediately. The species has been hunted extensively, leading to several population collapses until the first regulatory regimes were put in place, as we will see in Chap. 6.[28]

[26]Lavigne (2008), pp. 542–546.

[27]Kovacs (2015).

[28]Gentry (2008), pp. 788–791.

Over the last 40+ years, the population of the northern fur seal at the Pribilof Islands has declined by a further 66% despite the fact that they are now strictly protected under US law. Since this population constitutes around one half of the world's overall northern fur seal population, this decline is significant on a global scale. At the same time, the fur seal population at Bogoslof Island in the southern Bering Sea has grown at a rate of 48.5% between 1980 and 2005 and around 10% between 2005 and 2011. The overall population trend of this species is decreasing, but the increasing Bogoslof population, paired with a stable, if not increasing, population on the Russian islands, has counteracted this trend to a degree. Nevertheless, the IUCN lists the northern fur seal as 'vulnerable'. Climate change, pollution and human infrastructure—first and foremost for oil exploration, harbours and shipping—as well as entanglement in fishing gear constitute the major threats to this species. The small, local use for food does not impact the conservation status of the northern fur seal.[29]

2.2.3 Ringed Seal (*Pusa hispida*)

The ringed seal (*Pusa hispida*) is an earless seal species with a circumpolar distribution on the northern hemisphere. While the species is an inhabitant of salt water, two freshwater subspecies can be found in Lake Saimaa in southern Finland (Saimaa ringed seal, *Pusa hispida ssp. saimensis*) and in Lake Ladoga in southwestern Russia (Ladoga seal, *Pusa hispida ssp. ladogensis*).

The ringed seal is the smallest pinniped species, only reaching sizes of 1–1.5 m. Although mostly considered an inshore seal species with high population abundances close to land, ringed seals have also been observed on stable pack ice areas offshore and even at the North Pole. Their southernmost range, apart from the freshwater seals, extends as far south as Iceland and southern Greenland as well as into the Baltic Sea while most of the ringed seal population occurs north of the Arctic Circle.

Similar to harp seals, ringed seals spend considerable time in the water, but need solid ice for reproduction and breeding. Oftentimes, snow lairs and very remote locations also serve as haul-out sites for ringed seals. Therefore, population estimates are difficult to conduct.

Ringed seals are a seal species which spends considerable time underwater, but which maintains breathing holes in the ice, for which they scratch open the ice or build up snow with their fore flippers. Also snow lairs or caves are dug out by the species in which the seal rests during the winter and early spring. Ringed seal pups are usually born within these caves, that protect them from predators such as the polar bear.

Ringed seals constitute the main food source for polar bears and have been subject to subsistence hunts all over the Arctic for centuries. The catchy image of

[29]Gelatt et al. (2015).

a lonely Inuit hunter waiting by a breathing hole for a ringed seal to appear has found its way into public perception as a manifestation of subsistence sealing. Although an abundant seal species, its distance from the shore as well as the fact that throughout their life cycles they are rather solitary animals, commercial hunting has not reached the proportions of the harp seal, for instance. In Lakes Saimaa and Ladoga, ringed seals were considered a pest species in the past since they interfered with the fishing industry. To reduce seal populations, bounties were put in place by both the Finnish and Russian/Soviet governments.[30]

Since the collection of population data is unreliable, the overall population status of the ringed seal is unknown. Notwithstanding, the species is listed as 'least concern' since the overall population is estimated to exceed 3 million animals. While increasing in population size, the Saimaa ringed seal is highly endangered and has a total population of around 135–190 animals.. The Ladoga ringed seal displays an increasing population growth with an overall population of around 3000–4500 animals. It is listed as 'vulnerable.'[31]

2.3 Polar Bear (*Ursus maritimus*)

I think it's fair to say that the image of the lonely polar bear (*Ursus maritimus*) on a small ice floe is an image that we have all seen at least once in our lifetimes. Indeed, the polar bear is probably one of the most iconic marine mammal species—at least when it comes to demonstrating the adverse effect of climate change and the associated warming Arctic on the world's ecosystems.

Polar bears are on average the biggest of the bear species and occur only in the northern hemisphere. They are closely related to the brown (or grizzly) bear (*U. arctos*) and interbreeding occurs both in captivity and in the wild. Polar bears live in the North American Arctic, in Greenland and Svalbard, as well as in the Russian Arctic. Individual polar bears can also be found in Newfoundland or in Iceland. These are considered to be 'stray' polar bears and do not have their fixed habitat in these regions.

The interaction between humans and polar bears dates back several hundred years. As long as there have been people in the Arctic, polar bears have been hunted. In recent times, polar bears have shifted into the world's attention as an icon of climate change. The current state of public discourse oftentimes labels the polar bear as facing immediate extinction due to climate warming. However, the realities are somewhat more complex since there is not *one* polar bear population, but 19 in the polar bear range states. Of these, seven are indeed declining, four populations are stable, one is even increasing while the status of the remaining seven is unknown due to data deficiency.[32] However, media stories such as a polar bear 'invasion' of a

[30]Hammill (2008), pp. 972–974.

[31]Lowry (2016).

[32]Meltofte (2013), p. 115.

Russian town are easily linked with food shortages in the natural habitat and ultimately the effects of climate change.[33] While climate change is indisputable, we must also remember that humans encroach more and more onto polar bear territory. The availability of waste food and the associated presence of polar bears in human settlements is therefore also a means of the polar bear adapting to increasing human presence in its habitat.

Polar bears interbreed with brown bears, both in captivity and in the wild, referred to as hybridity. Although it has been known to occur for some time, it wasn't until 2006 when it was first observed in the wild when a sports hunter from the United States shot a hybrid in Canada's Northwest Territories, assuming it was a polar bear. Since then, several occurrences of 'grolar' (a mix of 'grizzly' and 'polar') bear sightings have been recorded. Whether or not the increasing existence of grolar bears is an effect of global warming is difficult to ascertain. On the one hand it is possible that due to the shrinking habitat of the polar bear, the habitats of grizzly and polar bears increasingly start to overlap, causing interbreeding. On the other, it is also possible that given the increasing human presence in the north, the existence of grolar bear is simply more often reported.

Another issue is that of genetic diversity. While indeed some polar bear populations are decreasing, the grolar bear could be a blessing and a curse. On the one hand, hybridisation can be a means to secure the genetic composition of the polar bear for the future. In other words, the polar bear would not genetically die out, but could still be found in the form of grolar bears. On the other hand, this could also be to the polar bear's detriment since the mix of both bear species is not fully adapted to either natural environment, ultimately speeding up the polar bear's decline.[34]

Be that as it may, since seven of the distinct polar bear populations are data deficient, it is not possible to generate reliable population estimates. The IUCN lists the polar bear as 'vulnerable' and the overall population is estimated somewhere between 19,000–30,000 bears. Whether or not the overall trend is increasing or declining is unknown.[35]

2.4 Summary and Conclusion

The above pages have introduced some of the most prominent marine mammal species, particularly taking into account public discourse and public perception. In short, the marine mammals presented in this chapter have been or are in some form relevant for humans, human perception, or human exploitation. This chapter is far from complete and entire books have been written about the diversity and roles of marine mammals for the ecosystem as well as for humans.

[33] Specktor (2019).

[34] Stirling (2008), pp. 888–890.

[35] Wiig et al. (2015).

A brief introduction to these species appeared necessary to better understand the role they play for humans and which legal responses have been taken to either regulate their hunts or to protect them. Upon reading the subsequent chapters, the reader is encouraged to either go back to these pages to recall which species we are dealing with, or to visit the website of the IUCN Red List of Threatened Species, available at www.iucnredlist.org, which holds a wealth of information on the status of the species in question, the history of human interaction with the species, and the current threats to the species. Probably the most authoritative source on marine mammals is the Encyclopedia of Marine Mammals[36] which is the most comprehensive volume on marine mammals so far.

Indeed, marine mammal biology is highly complex—both in regard to understanding the species and in regard to assessing the population statuses. For a better understanding of the benefits and pitfalls of different study techniques, I recommend the book *Marine Mammals. Biology and Conservation*, edited by Peter G.H. Evans and Juan Antonio Raga.[37] Also the website of the North Atlantic Marine Mammal Commission (NAMMCO) holds interesting information on both the conservation statuses of North Atlantic marine mammals as well as different survey techniques. The website is available at www.nammco.no.

Having obtained some knowledge on marine mammals, we now turn to the history of the interaction between humans and marine mammals.

References

Aguilar A (2008) Fin whale. In: Perrin W, Würsig B, Thewissen JGM (eds) Encyclopedia of marine mammals, 2nd edn. Academic Press, Burlington, pp 433–437

Ballance LT (2008) Cetacean ecology. In: Perrin W, Würsig B, Thewissen JGM (eds) Encyclopedia of marine mammals, 2nd edn. Academic Press, Burlington, pp 196–201

Berta A (2008) Pinniped evolution. In: Perrin W, Würsig B, Thewissen JGM (eds) Encyclopedia of marine mammals, 2nd edn. Academic Press, Burlington, pp 861–868

Clapham PJ (2008) Humpback whale. In: Perrin W, Würsig B, Thewissen JGM (eds) Encyclopedia of marine mammals, 2nd edn. Academic Press, Burlington, pp 582–585

Cooke JG (2018a) Balaenoptera musculus. The IUCN Red List of Threatened Species 2018. https://doi.org/10.2305/IUCN.UK.2018-2.RLTS.T2477A50226195.en. Accessed 16 May 2019

Cooke JG (2018b) Balaenoptera physalus. The IUCN Red List of Threatened Species 2018. https://doi.org/10.2305/IUCN.UK.2018-2.RLTS.T2478A50349982.en. Accessed 17 May 2019

Cooke JG (2018c) Megaptera novaeangliae. The IUCN Red List of Threatened Species 2018. https://doi.org/10.2305/IUCN.UK.2018-2.RLTS.T13006A50362794.en. Accessed 16 May 2019

Cooke JG (2018d) Eschrichtius robustus. The IUCN Red List of Threatened Species 2018. https://doi.org/10.2305/IUCN.UK.2018-2.RLTS.T8097A50353881.en. Accessed 20 May 2019

Cooke JG, Reeves R (2018) Balaena mysticetus. The IUCN Red List of Threatened Species 2018. https://doi.org/10.2305/IUCN.UK.2018-1.RLTS.T2467A50347659.en. Accessed 17 May 2019

36 Perrin et al. (2008).

37 Evans and Raga (2001).

Cooke JG, Zerbini AN (2018) Eubalaena australis. The IUCN Red List of Threatened Species 2018. https://doi.org/10.2305/IUCN.UK.2018-1.RLTS.T8153A50354147.en. Accessed 17 May 2019

Epstein C (2008) The power of words in international relations. Birth of an anti-whaling discourse. The MIT Press, Cambridge

Evans PGH, Raga JA (eds) (2001) Marine mammals. Biology and conservation. Kluwer Academic/Plenum, New York

Gelatt T, Ream R, Johnson D (2015) Callorhinus ursinus. The IUCN Red List of Threatened Species 2015. https://doi.org/10.2305/IUCN.UK.2015-4.RLTS.T3590A45224953.en. Accessed 21 May 2019

Gentry RL (2008) Northern fur seal. In: Perrin W, Würsig B, Thewissen JGM (eds) Encyclopedia of marine mammals, 2nd edn. Academic Press, Burlington, pp 788–791

Gillespie A (2005) Whaling diplomacy. Defining issues in international environmental law. Edgar Elgar, Cheltenham

Hammill MO (2008) Ringed seal. In: Perrin W, Würsig B, Thewissen JGM (eds) Encyclopedia of marine mammals, 2nd edn. Academic Press, Burlington, pp 972–974

Jones ML, Swartz SL (2008) Gray whale. In: Perrin W, Würsig B, Thewissen JGM (eds) Encyclopedia of marine mammals, 2nd edn. Academic Press, Burlington, pp 503–511

Jungh D, Sheldon KEW (2008) Bowhead whale. In: Perrin W, Würsig B, Thewissen JGM (eds) Encyclopedia of marine mammals, 2nd edn. Academic Press, Burlington, pp 131–133

Kalland A (2009) Unveiling the whale. Discourses on whales and whaling. Berghahn Books, New York

Kenney RD (2008) Right whale. In: Perrin W, Würsig B, Thewissen JGM (eds) Encyclopedia of marine mammals, 2nd edn. Academic Press, Burlington, pp 962–972

Kovacs KM (2015) Pagophilus groenlandicus. The IUCN Red List of Threatened Species 2015. https://doi.org/10.2305/IUCN.UK.2015-4.RLTS.T41671A45231087.en. Accessed 20 May 2019

Lavigne D (2008) Harp seal. In: Perrin W, Würsig B, Thewissen JGM (eds) Encyclopedia of marine mammals, 2nd edn. Academic Press, Burlington, pp 542–546

Lowry L (2016) Pusa hispida. The IUCN Red List of Threatened Species 2016. https://doi.org/10.2305/IUCN.UK.2016-1.RLTS.T41672A45231341.en. Accessed 21 May 2019

McGrath M (2018) Whale killing: DNA shows Iceland whale was rare hybrid. BBC News, 20 July 2018. https://www.bbc.com/news/science-environment-44809115. Accessed 16 May 2019

Meltofte H (ed) (2013) Arctic biodiversity assessment. Status and trends in Arctic biodiversity. Conservation of Arctic Flora and Fauna, Akureyri

Olson PA (2008) Pilot whales. In: Perrin W, Würsig B, Thewissen JGM (eds) Encyclopedia of marine mammals, 2nd edn. Academic Press, Burlington, pp 847–855

Perrin W, Würsig B, Thewissen JGM (eds) (2008) Encyclopedia of marine mammals, 2nd edn. Academic Press, Burlington

Sears R, Perrin WF (2008) Blue whale. In: Perrin W, Würsig B, Thewissen JGM (eds) Encyclopedia of marine mammals, 2nd edn. Academic Press, Burlington, pp 120–124

Sellheim N (2017) Japanese whalers and Canadian sealers – powerless under discourse and law? Curr Dev Arc Law 5:104–109

Specktor B (2019, 11 February) 52 Polar Bears 'Invade' a Russian Town to Eat Garbage Instead of Starve to Death. LiveScience. https://www.livescience.com/64741-polar-bears-are-taking-back-russia.html. Accessed 21 May 2019

Stirling I (2008) Polar bear. In: Perrin W, Würsig B, Thewissen JGM (eds) Encyclopedia of marine mammals, 2nd edn. Academic Press, Burlington, pp 888–890

Taylor BL, Baird R, Barlow J, Dawson SM, Ford J, Mead JG, Notarbartolo di Sciara G, Wade P, Pitman RL (2008) Physeter macrocephalus. The IUCN Red List of Threatened Species 2008. https://doi.org/10.2305/IUCN.UK.2008.RLTS.T41755A10554884.en. Accessed 20 May 2019

Wells RS, Scott MD (2008) Common bottlenose dolphin. In: Perrin W, Würsig B, Thewissen JGM (eds) Encyclopedia of marine mammals, 2nd edn. Academic Press, Burlington, pp 249–255

Wells RS, Natoli A, Braulik G (2019) Tursiops truncatus. The IUCN Red List of Threatened Species 2019. https://doi.org/10.2305/IUCN.UK.2019-1.RLTS.T22563A50377908.en. Accessed 20 May 2019
Whitehead H (2008) Sperm whale. In: Perrin W, Würsig B, Thewissen JGM (eds) Encyclopedia of marine mammals, 2nd edn. Academic Press, Burlington, pp 1091–1097
Wiig Ø, Amstrup S, Atwood T, Laidre K, Lunn N, Obbard M, Regehr E, Thiemann G (2015) Ursus maritimus. The IUCN Red List of Threatened Species 2015. https://doi.org/10.2305/IUCN.UK.2015-4.RLTS.T22823A14871490.en. Accessed 21 May 2019

3 Marine Mammals and Humans

3.1 Introduction

Ever since I saw my first whale, I have been fascinated by these animals. Technically speaking, it was a small cetacean, a harbour porpoise (*Phocoena phocoena*), that I saw lying dead on a beach in western Denmark. Where I grew up, in north central Germany, we certainly did not have any whales and therefore seeing one was a quite life-changing experience. This feeling grew stronger even when I saw my first common minke whale (*Balaenoptera acutorostrata*) and then my first humpback (*Megaptera novaeangliae*) in Iceland. After having seen these whales I immediately felt that these incredible creatures need to be protected. It was almost like my own epiphany. After all, the common discourse on whales and whaling has always contrasted their intellect, their social life and, indeed, musicality to their victimhood of commercial whale hunts, additionally affected by noise and other pollutants, being struck by ships and being caught as by-catch in fisheries. While, along with newer threats such as the ingestion of plastic, noise, pollution and ship strikes are undoubtedly major causes for the decline of whale populations, today, overhunting is arguably no longer such a cause. We will delve more into this issue below. So what is it that makes us focus so much on whaling and not other threats? Are all whales really intelligent, social and musical, almost comparable to humans? As we have seen in Chap. 2, this highly depends on the species and, concerning their population status, also on the subpopulation. The narrative of 'the whale' is, obviously, a gross oversimplification and neglects many biophysical features of different whale species. This predominantly Western depiction of whales was coined the 'superwhale' by the late Norwegian anthropologist Arne Kalland.[1] He claimed that without distinction, reference is made to this particular 'superwhale', making it impossible to justify any utilitarian aspects of different whale species in different regions of the world.

[1]Kalland (2009).

N. Sellheim, *International Marine Mammal Law*,
https://doi.org/10.1007/978-3-030-35268-4_3

When I was standing on a whale watching boat in Iceland, it is exactly these issues—whaling, whale intelligence, social behaviour, threats—that came to my mind. After all, in Iceland whales are hunted for commercial purposes, which means that primarily their meat is used as a commercial commodity. Whales are hunted to make money. This translates into whale meat not only being found on the shelves and in the freezers of supermarkets in Iceland, but also in restaurants and supermarkets in Japan—the main international customer of Icelandic whale meat. Whales are also hunted commercially in a different manner: the whale watching industry is booming, making Iceland follow a trend that has seen a tremendous growth of whale watching activities worldwide. Between 1998 and 2008 (the most recent data that I have seen), for example, the number of whale watching tourists had grown from 9 million to 13 million while its global revenue had doubled from US$ 1 billion to US$ 2.1 billion. All in all, 119 countries offer whale watching as part of their nature-based tourist industry. Apart from its economic value for a country's GDP, whale watching also offers socio-economic and educational benefits to the communities involved. Thus, economic activity and nature conservation can benefit each other.

Whale watching inevitably uses the awe-inspiring physique of whales—their aesthetic charisma—to influence the emotions of tourists. In Iceland, it is particularly minke and humpback whales that are being chased for photographs and life-long impressions. This commercial whale hunt is, by definition, non-extractive (leaving aside the extracted images and impressions) and thus non-lethal. However, this is by no means to say that whale watching is environmentally harmless. One must bear in mind that where there are tourists there is waste, noise and other impacts on the environment. In the context of whale watching, disturbance and ship strikes also belong to frequently occurring side-effects of this activity. In order to prevent increased disturbance by whale watchers, the IWC has developed the *Whalewatching Handbook.*[2] Whale watching is thus now fully integrated into the discourse on whales and human interaction with them. Indeed, one might even go so far as to say that whale watching is considered to be the future of human-cetacean interaction. Needless to say, this has not always been the case.

While many push for a non-extractive use of whales and for vegetarian and vegan diets, it is truly impossible to imagine Western society in which meat is fully abandoned. While there is a need to significantly reduce animal suffering (and to reduce one's environmental and carbon footprints), this wishful thinking contrasts with the harsh realities of life in industrial societies: they call for diverse and easy diets, neatly packed in plastic, always available all year round on the shelves of a supermarket. In practice this affects the way meat is being produced: in large-scale farms and slaughterhouses in which millions of animals are being bred and killed. The dimensions are truly staggering. Just take the poultry slaughterhouse in Wietze, Germany, for instance—the area where I grew up. In this slaughterhouse, the maximum killing capacity is 27,000 chickens per hour, amounting to 432,000

[2]IWC (2019).

chickens per day, more than 3 million per week and a mind-boggling 157,248,000 chickens per year!

Marine mammals have never seen such large numbers of killing. While certainly their size plays a role, they are, after all, animals living in the wild that humans need to physically hunt in order to consume them. Of course, this is not to say that marine mammals have not been bred in captivity: the polar bear *Knut* which was born in 2006 and died in 2011 at the Berlin Zoological Garden is but one example of a captivity-bred marine mammal. Further, this is not to say that marine mammals have never seen vast numbers of killing. Over the course of several centuries, millions and millions of seals and whales have been killed for commercial purposes which drastically reduced their numbers to the brink of collapse. In some cases, they have all but disappeared. For example, the last known specimen of the Caribbean monk seal (*Neomonachus tropicalis*) was sighted in 1952 between Jamaica and Nicaragua. Also during the so-called 'Whaling Olympics'—the decades after the Second World War until the adoption of the whaling moratorium—millions of whales were killed, driving particularly some large whale species close to extinction. At the same time, countless smaller cetaceans have been hunted in the drive fisheries of the Faroe Islands (*Grindadráp*) or in Japan (*oikomi-ryō*, 追い込み漁), to name a few.

In this chapter we get to know some of the cultures that have hunted and are still hunting marine mammals. Naturally, it is impossible to delve into *all* hunting cultures of the world since this would significantly go beyond the scope of this chapter. You will find literature in the 'Further Reading' list at the end of this chapter that both engage deeper with the socio-economic and cultural aspects of marine mammal hunts and which cover large geographical areas. In this chapter we focus on the geographical regions and countries that are still relevant in the discourse on marine mammal and marine mammal use: the Arctic, the Faroe Islands and Iceland, Newfoundland (Canada) and Japan. While we focus on these, you, the reader, will furthermore learn of whale hunts in the Caribbean and of whaling and sealing in Australia. The point of this chapter is to show that marine mammal utilisation has had a tremendous impact on peoples and cultures all over the world.

3.2 Marine Mammals and Human Societies

3.2.1 The Arctic

The dimensions of industrial marine mammal hunts are unimaginable in the frosty realms of the Arctic. Here, Arctic peoples have hunted seals and whales for generations—and indeed for millennia. While in the northern Atlantic it was primarily the Norse people in Norway, Denmark, Sweden and the islands of the North Atlantic who hunted seals, whales and walruses, the same can be said about the Inuit cultures of Greenland, Canada, Alaska and Chukotka. The importance of marine mammals is reflected in the myths and knowledge of the Arctic as well as and in unwritten, customary laws. Sharing of food or work, for instance, has been an integral part of Inuit societies, based on the availability and nature of abundant marine mammals. Also, one such unwritten law saw the immediate processing of seals by women. Otherwise seals would see a lack of availability during the subsequent hunt. While this, for the reader from an industrialised society, might appear somewhat strange, these kind of rules meant that (1) social structures were upheld; (2) the seal was processed right away and its products retained the highest quality; and (3) that men as the hunters of a community were able to hunt more and thus provide for the community instead of having to deal with the processing of the food.

The hunt for marine mammals in the Arctic existed before the emergence of contemporary Inuit — the Palaeo-Eskimo cultures. All along the shores of the Arctic seas already 4000—6000 years ago whales and seals were hunted and used. In excavated settlements of the Saqqaq people in Western Greenland, for instance, remains of numerous whale and seal species have been found and their age was traced back to around 2200 BC. Also the Dorset people that emerged in Eastern Canada around 2800 years ago and migrated to West Greenland, northern Newfoundland and Labrador used 'Arctic small tool technology'—highly specialised, blade-based tools—to hunt walrus and different seal species.

The continuity of the hunt for marine mammals in the Arctic is noteworthy—after all, it is still being pursued today. In Western Greenland, for example, which was populated by the Thule people near the turn of the first millennium, effectively replacing the Saqqaq, the remains from whale hunts date back to the same time period. The evidence point towards a long-standing whaling culture: permanent villages, indicating the potential for functioning whaling crews; boat parts and whaling gear; whale bones; remains from ritual practices; and remains from non-local goods traded from elsewhere that served as supplements to the local diet. In fact, the importance of marine mammals for the Inuit was that great that they are also reflected in myths of creation, which is worth citing in full:

> Long ago, there lived in the Arctic a young and beautiful woman. Many men were presented to her as prospective husbands, but she rejected them all, until one day a handsome stranger came to her family's camp. He had many dogs and finely crafted hunting tools, and he promised her a life of comfort. She could not resist, and she left with him for his land across the sea, only to find on her arrival that she had been deceived. Her betrothed was in fact a sea bird, and his dwelling a hovel. When her father came to rescue her, they escaped in his boat across the calm sea. But her seabird husband's spirit, angered by this, pursued them and whipped the icy sea into a raging storm. In midocean in a small boat, her father feared for his life, and, to appease the bird, cast his daughter overboard into the wave-tossed waters. But she clung relentlessly to the gunwale of his boat. Determined to release her grip, her father cut off her fingers at the first knuckle; the pieces fell into the sea and became the ringed seals. Still she held on. Then he cut off all her fingers at the second joint, and those pieces swam away as bearded seals. Other pieces of her fingers became walruses and whales. That is how seals were created, and that young woman became Nuliajuk, the Mother of the Sea... To this day she lives at the bottom of the sea and controls all the animals in her watery domain.[3]

[3]Reproduced in Stein and Stein (2016), p. 81.

Going even further, it is notable that in the language of the Inuit—*Inuktitut*—the similarity between 'meat' and 'eating' is striking: *niqi* versus *niri*. At the same time, contrary to the English language, for example, where there is only one word for 'fat', *Inuktitut* distinguishes between fat from the sea (*uksuk*) and fat from the land (*tunnuk*). Without further having to delve into the normative aspects of language it becomes clear that marine mammals have played a significant role in Inuit societies. It was such an important contributor to human survival in the harsh conditions of the Arctic, it also affected social and cultural practices. When the first Europeans entered the scene in the North and started to explore and exploit the Arctic, they were quickly faced with the difficult dietary needs human beings face in the cold. It soon became clear that sufficient provisions of fresh meat from seals and whales enabled overwintering and prevented the outbreak of scurvy. In other words, without meat human survival would not have been possible in the Arctic. Indeed, ethnographer Franz Boas went as far as to call seals the 'staple food' of the Inuit.[4] Its dietary importance is further underlined by a common depiction of the Inuit seal hunt: a hunter standing in a bent-over position over a breathing hole of a seal, harpoon in his hand, waiting for the seal to surface in order to strike it. This extremely uncomfortable position could last for hours, aggravated by cold winds and bad weather. It nevertheless underlines how vital seals were as a source of food.

In order to maximise their hunting efficiency, Inuit developed sophisticated technology in their pursuit of marine mammals. In seal hunting, for example, harpoons with detachable heads were used, linked to the hunter by a long string. The small, light boats from which seals were hunted in the sea are still in use today—*qajaq* or *kayaks*. This Inuit invention that is also part of our Western culture is a direct result of the hunts for marine mammals in the Arctic. In whale hunting, heavy harpoons, also with detachable heads, were thrust into the whale, which in turn was attached to several buoys made of inflated seal skins that prevented it from diving and escaping.

Social practices such as sharing, which can be found in indigenous societies all over the circumpolar North ensure the survival of all. Ongoing trade between different social and cultural groups enabled a variety in diet and goods, proving, once again, to be an important mechanism for human survival. In Chukotka—the far east of Russia—, for example, trade in goods was ongoing long before the region was colonised by Russians and coastal Chukchi and Yup'ik conducted trade with reindeer herding Chukchi of the inland and, in fact, all over the Arctic. Since certain seal species, as well as the walrus, followed the abundant sea ice and polynyas, seal and walrus hunters were forced to follow them. Settlement patterns were therefore not static but directly dependent on the availability of prey species. Marine mammals are also reflected in the cosmology of the Chukchi in which seals and walruses take a crucial place. Cosmological figures that are based on marine mammals, in turn, serve as names for places and children, once again underlining the normative importance of marine mammals in these societies. The normative role of the bowhead whale for

[4]Boas (1888/1964), p. 63.

Chukchi societies is a further testament to this claim: bowheads were intensively whaled between the ninth to the fifteenth centuries, reflected by large numbers of bowhead remains in ancient settlements, in oral traditions, rock paintings, sculptures and amulets. Even though only scattered records of whale festivals and rituals exist, Igor Krupnik concludes that "the symbolic role of the bowhead whale seems to have been extremely important in both Eskimo and Maritime Chukchi cultures. During specific periods, it reached a particularly high level of elaboration, as judged by complex ceremonial centers."[5]

Not surprisingly, polar bears have also been important for the inhabitants of the Arctic. Before contact with European colonisers and the associated trade in modern weaponry, the hunt for polar bears was an extremely dangerous endeavour. After all, polar bears are extremely defensive of their pups and when attacked. Consequently, hunting polar bears did not reach the same levels as seals or whales. Polar bears were only hunted to provide for extra food or clothing. During the hunts, extensive usage of sled dogs was made that encircled the bear until the hunter was in reach to kill it with spears or bow and arrow. Another technique was killing the bear during hibernation in its den before it awoke. However, polar bears lose significant amounts of body mass during the time of hibernation. Consequently, killing polar bears that were not in hibernation was, from a nutritional perspective, more efficient. Even though polar bears did not occupy the same place as whales and seals on a day-to-day basis, their role in the cosmology and religious practices in Inuit cultures is undisputed. Reference to polar bears or polar bear skins is a commonly recurring element of oral traditions.

3.2.2 The North Atlantic

It is easy to depict Old Norse as the brute seafarers, violently subjugating entire regions such as Normandy in France. Common representation of the 'Vikings' fuel this image although, to be fair, in recent years the representation of the ancient Norse has become more nuanced. After all, the Norse were not only warriors, but primarily skilled traders. Their trade network stretched from the North Atlantic, to Western Europe and into the contemporary Baltic states and probably even further towards the East along the main waterways. While marine mammals did not play an overarching role in the trade towards the East and West, they still constituted an important element of trade as well as a resource for settlement in the North Atlantic.

We must bear in mind that it was the Vikings from Denmark and Norway that landed on the shores of the Hebrides, Orkney, the Shetlands, the Faroe Islands, Iceland and, for a limited period of time, Greenland. Viking remains were even found on the northern tips of Newfoundland, Canada. In other words, Viking settlers landed all over the North Atlantic. The hunt for marine mammals played an integral role in the survival strategies of the Norse and archaeological evidence shows that

[5]Krupnik (1987), p. 21.

whale drives, particularly those concerning pilot whales, took place in the Hebrides, the Orkneys and most frequently in the Faroes. Occasional drives also occurred in the fjords of Norway, albeit on a much more opportunistic basis since the whales did not occur regularly in mainland Norway. This means that pilot whale driving techniques and technologies developed predominantly in the Faroe Islands during the time of settlement in the tenth century. The North Atlantic pilot whale drive is therefore a uniquely Faroese invention. Poor soil conditions that allow only for a very limited reliance on agriculture meant harvesting marine resources was crucial for the survival of the Faroese. Towards the end of the thirteenth century, Old Norse legal codes applicable in Norway, the Faroes and Iceland had established property regimes for whales. By then, whaling and whales had become an integral part of the Faroese economy.

Apart from the local utilisation of seal and whale resources, these also served as important trade goods for the trade with Europe, particularly England. Whale oil, salted and dried meat were commodities that the Norse traded in order to provide their settlements with other trade goods to diversify their diets and economies. Moreover, marine mammal products were used as luxury trade goods, especially walrus ivory. The supply of walrus ivory relied on two sources: one in the Barents Sea of northernmost Europe and northwestern Russia, and one on the west coast of Greenland. Walrus ivory which was recorded in important trading posts such as Novgorod, points towards the importance of this commodity in medieval Europe and Russia. The hunting grounds in the Barents Sea were first recorded during a visit of Norwegian chieftain Ohthere (Ottar) to King Alfred of Wessex in the ninth century.

Between the 1120s and the early fifteenth century, most of the European ivory originated in Greenland. In the times of significant economic growth in Europe between the eleventh to thirteenth century, also the demand for ivory grew. Consequently, given the predominant abundance of Greenlandic ivory in Europe during that time, it appears reasonable to assume that Greenlandic Norse settlements economically relied on this trade. In fact, the decline in the ivory trade, either due to overharvesting of walrus, climatic changes, outmigration, or declining demand in Europe may have significantly contributed to the collapse of the Greenlandic Norse settlements.[6]

Further south, in the waters of the Baltic and North Seas, it was mostly seals that were hunted. Even though seal hunts have been documented dating as far back as 43,000 years ago and as far south as Spain, it is along the Danish coast and into the reaches of the Gulf of Bothnia and Gulf of Finland that seal hunts constituted an integral part of the local and trade-based economy. During the time of the Holocene Climatic Optimum, a warm period 9000—5000 years ago, along the coasts of the Baltic Sea, settlements indicate that seals constituted a vital commodity for human survival. It is not surprising that in contemporary Sweden and Finland a unique terminology developed that related to the hunt for seals. Also in the Gulf of Finland, in Estonia, Latvia and Lithuania, seals were hunted. Trade occurred mostly during

[6]Star et al. (2018).

Fig. 3.1 Whale flensing in the Faroe Islands and seal hunting in the Gulf of Bothnia. Excerpts from Olaus Magnus, Carta Marina, 1539 (© public domain)

the heyday of the Hanseatic League from the fourteenth to the sixteenth century. However, it was not seals from the Baltic Sea that were sought after, but rather seals put on the market by Norwegian traders. The logs of the 'Bergenfahrer', those shipping from the German city of Lübeck to the Norwegian city of Bergen, indicate that Norwegian-produced seal blubber and fat were important trade goods for central Europe. The seminal map *Carta Marina* by Swedish cartographer Olaus Magnus from 1539 impressively shows how a whale is flensed on the Faroe Islands and how hunters harpoon seals in the Gulf of Bothnia (see Fig. 3.1).

This indicates that whaling and sealing in those regions were indeed part of the common discourse. While Olaus Magnus' map placed imaginary creatures and sea monsters on his map, it nevertheless also depicts contemporary understandings of the regions in question. The examples from the Faroe Islands and from the Gulf of Bothnia, paired with the archaeological and documentary evidence can be seen as real-life presentations of marine mammals and their use.

Human societies in the North Atlantic could not have developed without relying on the sea. While the Faroe Islands and Iceland nowadays are industrialised, 'modern' societies, large parts of their economies still depend on fisheries and other maritime activities. Of course, marine mammals also play a role. Most importantly, whales have provided the peoples of these remote islands with meat. Seals, on the other hand, have played a smaller role. In Iceland, seals have been consumed for human purposes, but have never been exploited commercially. Although their skins and oil had been a commercial commodity, potential commercial exploitation only started in the late nineteenth century. By then, however, the global markets for seal oil had already plummeted. Paired with the Great Depression of the 1920s as well as demographic shifts from smaller to larger settlements, commercial seal hunts did not take off. Seals are, after all, considered more of a nuisance animal than a commercial species. The fact that seals often carry the parasitic ring worm (*Pseudoterranova decipiens*) has made them unattractive as a staple. Since the ring worm affects fish stocks, the Icelandic government has been primarily interested in managing rather than commercially using seal populations since the 1980s.

Whales have provided the Icelandic people with significant amounts of food and economic income since the island's settlement. Therefore, Icelandic culture holds countless references to whales and whaling. Since 1948 it has been one company, *Hvalur H/F*, which has been operating land stations in Iceland, catching around 300 whales per year. When the moratorium on commercial whaling under the IWC came into effect in 1985/86, these whaling operations ceased since Iceland was a member of the Commission and did not object to the moratorium. In the same year, two whalers were sunk in Reykjavík harbour by members of the *Sea Shepherd Conservation Society* (SSCS). While Iceland conducted a limited degree of research whaling after the moratorium had come into force, it was pressured from other countries to halt this, despite scientific whaling being legal under the whaling convention (see Sect. 5.2.3). Due to its unwillingness to give up whaling, Iceland left the IWC in 1992, but rejoined 10 years later, this time with an official reservation towards the moratorium. In 2006, Iceland started hunting whales for commercial purposes again.

The Faroe Islands have seen whaling since their first settlement. Over the centuries, whaling has been an integral part of the islands' economic and socio-cultural systems. While in Iceland, over time, attention shifted to whales offshore, Faroese whaling has concentrated on the whale drive, meaning the driving of small cetaceans into the shores of the islands where they are killed. This is referred to as the *Grindadráp*. Even though the hunt is strictly regulated, it has been subject to intense international pressure. Organisations such as SSCS or *People for the Ethical Treatment of Animals* (PETA) have launched large international campaigns against the Faroe Islands to force it to stop its whale hunt by, for example, targeting the Faroese fish exports and lobbying governments to instigate a ban on Faroese fish products.

Seals have never been hunted to this extend. Mythologically, seals have been regarded as former humans that have sought death in the sea, but that also return back to land. One of the most popular tales in the Faroes is that of the *Kópakonan*, the Seal Woman. The story manifests a long-standing perception of seals in the North Atlantic: the selkie—a hybrid creature between seal and human. Since, contrary to whales, seals spend their lives on both land/ice and in the sea, mythologies of the North Atlantic treat them accordingly. Icelandic anthropologist Níels Einarsson refers to them as "betwixt and between' in the land/sea dichotomy."[7] While this was also the case for Norway, the Norwegian relationship with seals has been marked by both traditional selkie-perception and as a commercial commodity, as we have seen above. While this accounts for the ethnic Norwegians, this is also true for the indigenous Sámi whose mythology is characterised by the Arctic environment. Stretching across the Arctic regions of four countries in northern Europe—Norway, Sweden, Finland and Russia—the Sámi are commonly known

[7]Einarsson (1990), p. 37.

as reindeer herders. For many, however, primary livelihoods were based on the sea. Seal products can still be found on Sámi markets all over the traditional Sámi homeland.

Norway has also been a major player in the international hunt for seals in other regions of the North Atlantic. It has conducted sealing predominantly at Jan Mayen Island east of Greenland and in the Canadian Gulf of St Lawrence. Similarly, Norwegian whaling has been an activity all over the North Atlantic. Although this is the case, Norway is a member of the International Whaling Commission. Since it has formally objected to the moratorium on commercial whaling, it is not bound by it.

3.2.3 Atlantic Canada

Atlantic Canada is part of the North Atlantic region and has become synonymous for seal hunting. The island of Newfoundland is probably the most well-known region for the seal hunt. Countless images of skinned seals, videos of brutally slaughtered pups, and a discourse on cruelty, lack of necessity and ruthless exploitation has shaped the view of many when linking Newfoundland to hunting seals. And indeed, Newfoundland, part of the province of Newfoundland & Labrador, still sees ongoing seal hunts today.

While these are large-scale hunts, it can be easily forgotten that sealing has been a part of Newfoundland's history since the very early days of human settlement. When the first indigenous peoples settled in Newfoundland and elsewhere in Atlantic Canada, seals were hunted, used and consumed. It is therefore by no means a recent activity. When European colonisation occurred, seals were hunted to sustain the early settlements that sprung up as part of the exploitation of the abundant cod stocks for the European markets. Seals were so abundant that some time in the eighteenth century they were found to be a cheap alternative to whales since also seals could be used to produce oil. In 1793 it is reported that the first vessel left Newfoundland exclusively for the pursuit of seals. From then on, seals became one of the most important mainstays of the newly emerging economy of Newfoundland, leading to extremely large numbers of seals being taken during the nineteenth century. With the discovery of hydrocarbon fuels, the markets for seals (and whales) collapsed.

Apart from the economic value, are an integral part of Atlantic Canadian culture and identity. All over Newfoundland and the French-speaking Magdalen Islands in Québec, references to seals and the seal hunt can be found while a rich repertoire of seal hunt-related songs and poems exists.[8] Similarly, seal meat is still a common feature in the cuisine of Atlantic Canada: the traditional seal flipper pie as well as a modern, exquisite culinary landscape are prevalent. Moreover, for some communities along the coast, the sea with all its living resources has become an integral part of identity. The pride and awe to be part of a sea-faring social system

[8]Ryan and Small (1978).

can be directly linked to the seal hunt. Without it, an important signifier of Newfoundland and Magdalen Island identity would be lost.[9]

The Right Whale: How Species Conservation and Fisheries Collide

The North Atlantic right whale is a highly endangered whale species with an estimated abundance of 411 individuals (June 2019). It has been put under strict protection in the United States and in Canada. One of the major threats are ship strikes since major shipping lanes and the routes of right whales intersect. Between 1980 and 2010 conservation initiatives focused primarily on ship strikes. While this proved to be successful, another major threat—entanglement—was left unattended.

Some of the right whale's critical habitats in the Canadian Gulf of St Lawrence are located in areas which are particularly important for the local lobster and crab fisheries. An increasing number of right whales has been entangled in fishing gear, having led to population downturns. In May 2018 the Canadian government announced the closure of six lobster and crab fishing areas since right whales were sighted in them. This led to protests from the Maritime Fisheries Union which represents fishers and plant workers since the entire economy of several communities would be severely affected by a closure. As a compromise, the Union proposed refraining from closing shallower waters where right whales would not go anyway. This, however, was rejected by the government.

In June 2019, when these words were written, the Canadian government is following a multi-tier approach to right whale conservation: (1) shallow waters (up to 20 fathoms = 36.6 m) will be temporarily closed only when right whales have been sighted in them; (2) temporary closures of larger, deeper areas of lobster and crab fisheries will be undertaken when right whales have been spotted; (3) season-long closures will be undertaken in areas where 90% of sightings have occurred.

While this approach is to some degree a compromise, lobster and crab fishers are discontent since temporary and season-long closures mean significant losses for them. A solution that satisfies both fishers and whales is not in place at the time of writing. However, work on rope-less, acoustic trap identifiers is ongoing, which could provide for a long-term improvement of the situation.

The history and cultural importance of whaling in Atlantic Canada is different. Along the Strait of Belle Isle between Newfoundland and Labrador (mainland Canada), early Basque whalers established their first whaling stations in the sixteenth century. This open-boat, traditional whaling continued up until the nineteenth century and was dominated by different European powers.

[9] Sellheim (2015).

When the Norwegian Svend Føyn developed the explosive harpoon in the early second half of the nineteenth century, commercial whaling all over the world was transformed radically. While already many species such as the right whale had been heavily depleted, other species which had been more difficult to hunt were also decimated in pelagic sealing. Modern technology and the high yields meant that in Newfoundland shore-whaling stations were established. Yet with declining populations and the moratorium of 1982 (after which Canada left the IWC), whaling in Newfoundland declined and has not been relaunched.

While in Atlantic Canada is still the place for large-scale sealing, international campaigns have led to severely declining international markets for seal products. The closure of the markets in the European Union sent a ripple effect throughout the world (see Sect. 6.6). However, those seal processing facilities still operating now provide products for markets in Asia as well as in Canada domestically.

3.2.4 Japan

Next to the Canadian commercial seal hunt, Japanese whaling is probably the most well-known marine mammal hunt in the world. Japan's position within the IWC as an avid pro-whaling nation and ultimately its withdrawal from the organisation have placed Japan into the centre of international opposition. Japan's whaling has oftentimes been labelled as outdated, obsolete, without cultural or social backing and so forth. However, in the IWC, the Japanese government has repeatedly tried to place whaling in Japan within a discourse on culture and cultural need. The failed attempts to obtain recognition of Small-Type Coastal Whaling (STCW) as a new category similar to Aboriginal Subsistence Whaling within the IWC are but one example for Japan's tenacity on the role of whaling in Japan.

Leaving aside any emotional responses to the contemporary whaling issue, it is clear that along the coasts of Japan, whaling has for many centuries played an important role for local communities. While coastal whaling is restricted to merely a handful of small communities, in the past, whales were an important contributor to food security all over the country. While that is so, in a fascinating study, Mayumi Itoh has shown that whales have also elements in cultural activities: mourning ceremonies for whales were held since they were seen gods of the sea due to their appearance and their contributions to local economies and food security.[10] For example, Shinto shrines celebrating the god of fishermen and luck, Ebisu (恵比須), also were used to celebrate dead whales that were washed ashore as manifestations of Ebisu. The Nishinomiya Shrine (西宮神社) in Hyōgo Prefecture, for instance, is one of the primary shrines for the celebration of Ebisu.

In Taiji, where the dolphin drive takes place, every year, the Taiji Whale Festival (太地浦くじら祭) celebrates both the contributions of the whale as well as depicting the struggle between a large whale and small whaling boats. Wakayama

[10]Itoh (2018).

Fig. 3.2 Taiji rock engraving © Nikolas Sellheim

Prefecture has designated this 'Whale Dance' as an intangible cultural heritage.[11] All over the community, references to whales and whaling can be found. While some are of a rather recent origin—arguably put in place to substantiate Taiji's cultural whaling history—some are much older. For example, a rock engraving on the rocky shores of Taiji which refers to a man's responsibility for whaling boats is several hundred years old (Fig. 3.2).

Taiji is, naturally, not the only place in Japan where these relics can be found. All over the country, apart from intangible relics such as songs or dances, also monuments, shrines, temples and graves still exist, which are related in one way or another to whales and whaling. The commercial aspect, however, is a different matter and here, it is my belief, lies the crux of the issue. While whale and whaling relics are spread all over the country, Itoh's study also shows that one needs to actively look for them. From my own experience, the same can be said about whale meat: while it can be obtained in many places, one does indeed need to actively look

[11] Anon (2015).

Fig. 3.3 Whale meat booth in Osaka, © Nikolas Sellheim

for it. One such example is a booth at a market in Osaka, which exclusively sells whale meat (see Fig. 3.3).

Whether or not whale meat can be considered integral to Japanese cuisine is probably best left to the Japanese to decide for themselves. Many Japanese of around 40 or more years of age that I personally have spoken with about this issue note that whale meat used to be a common dish in schools. This was also due to the fact that after the Second World War, when the country was economically on its knees, the American forces encouraged the introduction of whale meat as a staple in order to alleviate the food shortage the country was experiencing. It was only after the war that whale became a country-wide source of food—primarily as canned meat. Whether or not this is a matter of culture or cultural history remains for the people of Japan to decide for themselves and I refrain from making a judgement on this issue. Either way, anthropological studies have shown that whales and whaling have been a part of the socio-economic landscape in certain regions. The claim that Japan does *not* have a whaling history and that whaling is first and foremost a nationalistic power play[12] is factually incorrect.[13]

[12]Morikawa (2013).

[13]E.g. Kishigami et al. (2013).

Whales and whaling are not the only marine mammal related activity in Japan. A somewhat lesser known history relates to the hunt for seals. Seal hunts have played a role in northern Japan, especially in Hokkaido. Here, the indigenous Ainu as well as Japanese settlers consistently made use of the marine fauna passing along it shores, including several seal species. In Ainu culture, the seal hunt left cultural traits such as rituals on the departure of the soul from a dead animal, which included seals. While these traits do exist, these are not as plentiful as, for example, whales and whaling in Japanese society. What can be said, however, is that seal skins were important trade goods between the Ainu of Hokkaido and Sakhalin with the Japanese, at least since the seventeenth century. When the fur seal herds of the North Pacific were discovered as a commercial commodity, many Ainu became part of their large scale exploitation—both voluntarily and involuntary. Moreover, northern Japan served as an important hub for foreign vessels aiming to reach North Pacific waters to hunt seals. Nowadays, however, seals are no longer exploited commercially. Those seals that can be found along Japanese shores are hunted due to their negative impact on salmon fisheries. It is meat from these hunts that is sold in limited amounts in northern Japanese supermarkets.

3.2.5 Other Regions and Countries

Contemporary discussions around marine mammal hunts oftentimes focus on current whaling and sealing activities and specifically those regions or countries in which either commercial hunts (e.g. Canada, Japan, Norway, Iceland) or subsistence hunts (e.g. by the Inuit) are conducted. It is easily forgotten that both whaling and sealing have a long-standing history all over the world. Some countries that have a whaling and sealing history are now outspoken opponents of any marine mammal hunts. In Australia, for example, there is a rich history of whaling and sealing. In fact, sealing was so important to the settlement of Tasmania that even up to this day the demography of the island is an effect of sealing and associated settlements. While colonial sealers forcefully subjugated the local aboriginal population and exploited their sealing skills, they paradoxically are also responsible for the preservation of aboriginal cultures and languages: through marriage with aboriginal women and the development of creole languages and practices the total colonial eradication of aboriginal culture was prevented. The effects were so long-lasting that Lynette Russel even goes so far as to say that these mixed communities have provided the basis for the contemporary Aboriginal community in Tasmania.[14]

Australia also has had a long-standing whaling history. Like sealing, whaling was an important industry when Australia was colonised. Exports to Europe and Asia provided the young colony with the revenue to establish sustainable industries and economies. Some time from the eighteenth century until 1978, Australia had been a whaling nation. Just to name two examples out of many all over Australia: on the one

[14]Russell (2012), p. 113.

hand, Australian whalers pursued whales in Antarctic waters. On the other, in Western Australia, shore-station whaling was conducted by French and American (Yankee) whalers. In the south of Western Australia, in Geographe Bay, the Castle Bay Whaling Station, which operated between 1845 and 1872, is directly related to the foundation of the town of Dunsborough. After all, Geographe Bay is a region from which whales can still be seen from land (see Fig. 3.4).

Australia's own experience with whaling and the detrimental effects of other commercial whaling on whale populations have made the country a leading country against any form of commercial whaling. It even went so far as to trigger legal proceedings against Japan before the International Court of Justice due to Japan's whaling activities in the Antarctic (see Sect. 5.3.5). This being said, Australia has, at least to my knowledge, not opposed marine mammal hunts *per se*. For example, it has always supported aboriginal whaling while the culling of seals in Tasmania due to their effects on the fisheries sector are considered. Nevertheless, Australia has often clashed with other nations, for example within the IWC, for its rigid stance against commercial whaling.

One outspoken pro-whaling country is St Vincent & the Grenadines, where also aboriginal whaling of a handful of whales per year is conducted. This whaling takes place on the island of Bequia and is a direct result of the lack of agricultural yield to feed the approximately 5000 inhabitants. St Vincent & the Grenadines joined the IWC in 1981. Since Bequia whaling predominantly focuses on humpback whales, the adoption of the moratorium on commercial whaling in 1982, also affected this whale hunt. It was only in 2013 that a quota of 4 humpbacks was assigned under the Aboriginal Subsistence Whaling scheme. Since only traditional technologies such as hand harpoons and open boats can be used to hunt these whales, the quota is hardly ever met. Apart from humpbacks, small cetaceans such as orcas or dolphins are also hunted in St Vincent & the Grenadines. When whales were plentiful in the waters around Bequia, at its peak in the late nineteenth century, five whaling stations were operated by Americans and Europeans. Especially Yankee whaling for baleen whales embossed the hunting operations in Bequia. Still, the traditionality of Bequia whaling was accepted by the IWC and does not seem to be contested by Commission members.

3.3 The Commercial and Non-commercial Use of Marine Mammals

The discussion above has shown that there are essentially two motivations for hunting marine mammals: as a commercial commodity and as a means to fulfil immediate dietary and other needs (subsistence). In the past, marine mammals were considered a commercial commodity. This perception resulted in unsustainable hunts that resulted in population declines and that has shaped the common narrative of environmental degradation and cruelty. Questions remain, however, how 'commerce' and 'subsistence' are actually defined and in how far these are shaped by idealisations, demonisations and simplified understandings of hunts of the past.

Fig. 3.4 Whales of Geographe Bay, Australia © Nikolas Sellheim

Looking at these historical hunts, the line between commercial and non-commercial subsistence hunts could easily be drawn. The seal hunts in the Bering Sea as well as in eastern Canada and in the Arctic serve as representative examples. Bering Sea and Newfoundland seal hunts were driven by international market demands. This means that it was the demands of people elsewhere in the world that motivated the seal hunts. The conduct of these hunts served no other reason than to fulfil these demands, playing into the pockets of large businesses who aimed to fill these as quickly as possible by killing as many seals as possible in the shortest possible period of time. We have to remember that the peaks of these extremely large-scale hunts occurred when (1) modern technologies such as high-powered rifles, hydrocarbon-fuelled engines or industrial processing methods found their way into contemporary societies; (2) the industrial revolution was in large demand of fuel; (3) environmental considerations and knowledge were little to non-existent; and (4) a regulatory framework that aimed to protect species (and people) were either non-existent or only emerging. The keywords here are therefore 'industry' and 'markets' as prime motivators. Aboriginal sealing, be it in the Bering Sea or in Arctic Canada, played hardly any role in these developments. Therefore, its impact on the dwindling seal populations was negligible.

In contemporary seal hunts, this distinction cannot be drawn that easily anymore. Inuit sealers all over the Arctic are fully integrated into market-based economies and everything they do has been influenced by modern-day technologies and socio-economic mechanisms. In the case of the Inuit, for example, the sale of seal skins and other seal products is in fact a sale for domestic and international markets. Currently, the subsistence seal hunt in Greenland numerically trumps the commercial seal hunt in Canada. Either way, both in Greenland and in Canada, much of the seal is also used for subsistence purposes. In one way or another, money is involved and money earned from the sale of seal products oftentimes directly flows back into fuel and other goods so that it is possible to conduct sealing. After all, in modern-day subsistence hunts, Inuit sealers rely on modern technologies such as snow-mobiles or rifles. With the emergence and, arguably, ultimate success of the anti-sealing movement in the 1960s, buyers abstained from buying any seal products, irrespective of whether they were hunted by Inuit or not. This led to the collapse of most markets for seal products, heavily affecting Inuit as well. Although bodies like the European Communities and their seal products ban in 1983 aimed to exclude Inuit from any ban, the overall markets for seal products collapsed. Inevitably, Inuit sealing experienced a massive drop, leading to increased suicide rates and other signs of social instability.

This example shows how intertwined the pathways of commercial and subsistence sealing are. In contemporary Canadian sealing, seal products from the Arctic use the same trade routes as seal products from commercial hunts in Atlantic Canada. The more the markets for seal products are in decline, the more Inuit sealing is affected, too. This is why Inuit and non-Inuit sealers have formed alliances to revitalise international markets for seal products.

One of the main anti-sealing actors in this was *Greenpeace*. During the 1970s, the organisation massively campaigned and lobbied against Newfoundland sealing, yet it has always indicated that it would support subsistence sealing by Inuit. Unfortunately, the impacts of collapsing markets also hurt Inuit communities, resulting in *Greenpeace* being considered an adversary of the Inuit. While other organisations have continued their campaigns against sealing, in 2014 *Greenpeace* issued a formal apology to the Inuit, aiming for more cooperation especially in light of the environmental challenges at hand.[15]

What is often ignored in this context are the impacts of anti-sealing campaigns and market collapses on commercial sealers and their families. When I did my fieldwork in 2013 in Newfoundland, I participated in both the seal hunt and the seal processing industry. I was the first European outsider to gain insight into the industry and the effects collapsing markets had and have on community cohesion and social wellbeing. Apart from financial issues, many Newfoundlanders can still closely relate to the seal hunt since it has been an integral part of Newfoundland society and economy for several hundred years. The pride of working in an industry that is vilified by outsiders is met with a complete lack of understanding since, for example, 'the Europeans' obviously don't know what sealing means in practice, the dangers that are associated with it and the food it brings to the table.

Having participated in a sealing trip myself I was able to witness these dangers first hand: cold, darkness, ice, sharp knives and gaffs, waves, wind, slippery decks and extreme exhaustion after long days of hunting seals. The bonds that are forged in these situations last for a lifetime and are crucial to understand social dynamics in sealing communities. In the processing sector, the pride of working in an industry, which is driven by (mostly) men and which has such an important history in Newfoundland society is an important part of the workers' identity. With market collapse and associated industry decline, this deeply-rooted cultural feeling is lost. While on the surface the contemporary seal hunt in Newfoundland is commercial—it aims at the skin and blubber of seals for national and international markets—Newfoundland sealers interpreted it as a subsistence hunt since many products such as flippers or hearts are bartered or given. The motivator to hunt seals as a food source (either through monetary income or through meat on the table) steers this interpretation. Be that as it may, also with declining markets, seals are hunted in Newfoundland. While their number may go down, these so-called 'landman hunts' are carried out to obtain food without a market dimension.

As a last comment on this issue: in the twentieth century, Newfoundland sealing was carried out by large vessels carrying hundreds of sealers, operated by large commercial companies. This seal hunt was indeed commercial in nature. Nowadays, the seal hunts are carried out by small boats (up to 10 people), all operated by small businesses. The seal hunt is therefore not carried out to make profit, but rather to make a living.[16]

Commercial and subsistence whaling are also two seemingly confrontational concepts. An interesting case in point in this regard is the dolphin and whale drive

[15]Kerr (2014).

[16]Sellheim (2015).

in Taiji. There, only 20 whalers constitute the group of people hunting the whales. While this might indicate that whaling only benefits these few, in this case whale meat provides an important food source for the entire community. Most enlightening were my interviews with dolphin trainers in the aquarium associated with the Taiji Whale Museum. The predominantly young trainers feel an immense pride to be able to work with dolphins and whales as being of such importance for the Japanese people. The question of captivity did not occur to them in the same way as it would occur to people in, say, European societies. The approach to the environment, animals and animal welfare appears to be philosophically different than in the West.[17] It seems that it is the contribution of whaling and captive whales to Japanese society en gros which is the main source for the immense pride the trainers feel.

Once again, the line between commercial and subsistence appears somewhat blurry. Commercial aspects, such as the sale of live dolphins to aquariums, certainly plays a major role in the motivation. But to say it is only about the money oversimplifies the long history of whaling, the cultural aspects associated with it and, quite simply, the food that it provides for the people of Taij. The use of small cetaceans is therefore not merely driven by monetary considerations, but has a component which runs deeper.

The question that always arises in the context of marine mammal use is whether there is a need for their utilisation. In other words: can people live without it? From my own perspective I would claim that, for instance, Newfoundlanders or the people of Taiji *could* live without sealing or the dolphin drive. But why *would* they? The answer to that question can probably best be found in George Orwell's *Animal Farm*: "All animals are equal, but some animals are more equal than others."[18] This is to say that for some animal species, certain ethical and moral rules appear to apply. This is so in the case of marine mammals which are discursively removed from any considerations as a food source. In a recent report by the Food and Agriculture Organization of the United Nations (FAO) on microplastics in fisheries and aquaculture this was best exemplified: in a diagram showcasing the pathways of microplastics from marine animals to the human, the depicted whale is not connected to the human.[19] In other words, whales are not considered within contexts of food and microplastics. But as we see, in some regions of this world, marine mammals are still sources of food. There is therefore a nutritional need for them, at least from the perspective of the locals. Once again, I refrain from making a judgement on whether this perspective is justified or not. For it is my conviction that a 'need' cannot be determined by outsiders, but should be considered from a vantage point of local history, culture and socio-economic circumstances.

[17]Kagawa-Fox (2012).

[18]Orwell (1945/1951), p. 90.

[19]FAO (2017), p. 45.

3.4 Summary and Conclusion

The above has shown that marine mammals have been important contributors to human societies since time immemorial. While I have only used pinpointed examples, wherever marine mammals and humans have interacted, marine mammals have been hunted. During the time of the industrial revolution and during the twentieth century, this led to overhunting. Although driven by economic greed, the commercial and non-commercial hunts for marine mammals are important markers of local identity and history. This does not only count for aboriginal hunters and their deeply enshrined cultural history, but also for whalers and sealers that are considered 'commercial' under contemporary international discourse. Whether, from the outside, commercial marine mammal hunts can be considered within contexts of identity and culture is up to you, the reader, to decide. What I hope to have achieved with this chapter is to show that marine mammal hunting is a complex matter with significant social and historical dimensions that are often ignored.

Since this book is merely an introduction to the topic, this chapter is far from a complete ethnography on marine mammal cultures. If you wish to know more about the cultural dimensions, I encourage you to delve into the depths of the literature, some of which I have either directly referenced to or which I provide in the 'Further Reading' section.

References

Anonymous (2015) Taiji's cultural heritage. City of Taiji, Taiji

Boas F (1888/1964) The central Eskimo. University of Nebraska Press, Lincoln

Einarsson N (1990) Of seals and souls. Changes in the position of seals in the world view of Icelandic small-scale fishermen. Mar Anthro Stud 3(2):35–48

FAO (2017) Microplastics in fisheries and aquaculture. Status of knowledge on their occurrence and implications for aquatic organisms and food safety. FAO, Rome. http://www.fao.org/3/a-i7677e.pdf

Itoh M (2018) The Japanese culture of mourning whales. Whale graves and memorial monuments in Japan. Palgrave Macmillan, Singapore

IWC (2019) Whale Watching Handbook. https://wwhandbook.iwc.int/en/

Kagawa-Fox M (2012) Japan's global environmental policy. The conflict between principles and practice. Routledge, Abingdon

Kalland A (2009) Unveiling the whale. Discourses on whales and whaling. Berghahn Books, New York

Kerr J (2014) Greenpeace apology to Inuit for impacts of seal campaign. https://www.greenpeace.org/canada/en/story/5473/greenpeace-apology-to-inuit-for-impacts-of-seal-campaign/

Kishigami N, Hamagushu H, Savelle JM (eds) (2013) Anthropological studies of whaling. National Museum of Ethnology, Osaka

Krupnik II (1987) The bowhead vs. the Gray whale in Chukotkan aboriginal whaling. Arctic 40 (1):16–32

Morikawa J (2013) Whaling in Japan. Power, politics and diplomacy. Hurst & Company, London

Orwell G (1945/1951) Animal farm. Penguin Books, London

Russell L (2012) Roving mariners: Australian aboriginal whalers and sealers in the southern oceans, 1790–1870. State University of New York Press, New York

Ryan S, Small L (1978) Haulin' rope and gaff—songs and poetry in the history of the Newfoundland seal fishery. Breakwater Books, St. John's

Sellheim N (2015) Morality, practice and economy in a commercial sealing community. Arc Anthro 52(1):71–90

Star B, Barrett JH, Gondek AT, Boessenkool S (2018) Ancient DNA reveals the chronology of walrus ivory trade from Norse Greenland. Proc RSB 285. https://doi.org/10.1098/rspb.2018.0978

Stein RL, Stein P (2016) An anthropology of religion, magic and witchcraft. Routledge, New York

Further Reading

Fielding R (2018) The wake of the whale. Hunter societies in the Caribbean and North Atlantic. Oxford University Press, Oxford

Kalland A, Sejersen F (2005) Marine mammals and northern cultures. CCI Press and NAMMCO, Edmonton and Tromsø

Monks G (ed) (2005) The exploitation and cultural importance of sea mammals. Oxbow Books, Oxford

Sanger CW (2016) Scottish Arctic whaling. Birlinn Ltd, Edinburgh

Watanabe H (2009) Japan's whaling. The politics of culture in historical perspective. Trans Pacific Press, Melbourne

The Global Framework for the Environment and Marine Mammals

4

4.1 Introduction

In this chapter we will explore how the international community has been dealing with marine mammals and what multilateral regimes of relevance for marine mammal conservation exist. Broadly speaking, these fall within the realm of international environmental law and, to a lesser degree, the law of the sea. This is to say that their primary objective is the conservation (and from the outset sustainable use) of biodiversity through the protection and management of specific species. The Convention on International Trade in Endangered Species of Wild Fauna and Flora (CITES) from 1973, the Convention on Migratory Species of Wild Animals (CMS) or the Convention on the Conservation of European Wildlife and Habitats (Bern Convention) both from 1979 are such examples. Moreover, we will get to know the Law of the Sea Convention, which treats marine mammals in a special manner.

Therefore, the following pages introduce the global framework for the conservation of the environment and the role marine mammals play in these regimes. While we will get to know several global agreements for the regulation of the global commons as well as for the conservation and sustainable use of biodiversity, we will also delve into regional agreements in which marine mammals play a role. Lastly, we look into a common strategy for marine conservation: the establishment of marine protected areas (MPA) and their (potential) efficiency for conservation and their (potential) impact on marine resource use.

4.2 Global Regimes of Relevance

Marine mammals have for more than a century played a role in international cooperation, particularly in regard to seals (see Chap. 6). But apart from this species-based approach, international cooperation has furthermore considered marine mammals in other regimes, particularly following the emergence of the global environmental movement during the 1960s. It therefore does not come as a

N. Sellheim, *International Marine Mammal Law*,
https://doi.org/10.1007/978-3-030-35268-4_4

surprise that the first global conference on the environment, the UN Conference on the Human Environment (UNCHE), held in Stockholm in 1972 started to address marine mammal issues. Particularly the call for a moratorium on commercial whaling was uttered at the conference and logged as an official recommendation emerging from that meeting. While it was still to take 10 years until this call would become reality, the overall no-use discourse on marine mammals was in full swing. Inevitably, this would also affect the negotiations at the UN law of the sea conference, which started in 1973. "All animals are equal, but some animals are more equal than others"[1] would also find its way into multilateral environmental agreements and their relationship to marine mammals.

4.2.1 Trade in Wildlife

Apart from regulating hunts for species, their protection occurs also through the control of international trade or the imposition of trade barriers. So far, several times throughout this book the Convention on International Trade in Endangered Species of Wild Fauna and Flora (CITES)[2] has been mentioned. CITES is a global regime that was concluded over the assumption that the control of international trade of species that are endangered would positively benefit their conservation status. After all, this would mean that international trade in wildlife or wildlife products constitutes a major threat to a specific species. Or to turn the argument around, if international trade in a species were to be suspended, the conservation status of a species would benefit, even in light of other threats. I will not delve into this matter here, but suffice it to say that the efficacy of trade limitations is subject to debate.[3]

Irrespective of this debate, it is clear that most countries are parties to CITES. The decisions taken by this convention are binding upon the vast majority of states—183 in 2019. This is certainly also relevant for the international trade in marine mammals. Let us briefly consider how CITES works. First of all, CITES does not regulate trade in *all* wildlife, but only for specifically listed species. These species are listed in three Appendices: Appendix I prohibits all trade; Appendix II limits trade; and Appendix III requires international cooperation for trade in species. The parties to the convention meet every 2–3 years and debate further course of action, the listing of specific species on one of the Appendices, the down—or uplisting of species. These are not uncontroversial but accompanied by heated debate. While the convention is global in nature, a high degree of sovereignty rests on its members, which fuels this debate. This is best exemplified by the role science plays in the CITES working structure: while in principle all decisions are to be based on scientific advice, this advice is to come from national scientific bodies, thus leaving it to the respective national

[1] Orwell (1945/1951), p. 90.

[2] Convention on International Trade in Endangered Species of Wild Fauna and Flora, 3 Mar 1973 (993 UNTS 243).

[3] See for instance Couzens (2013), pp. 126–130.

government to take action (or not) and to interpret scientific findings. While the CITES Secretariat itself also conducts and oversees scientific issues, it itself cannot take decisive action, but rather recommends certain steps to the parties.

Once a two-third majority of the parties accepts changes to the Appendices, these become binding to all parties. However, parties do have the possibility to lodge formal reservations to the listing of specific species. Once the Secretariat has received these reservations, and until it has been formally withdrawn, the party with a reservation is not bound to the trade limitations accompanying Appendix-listings. This means that a particular party can still conduct trade in listed species with countries not party to CITES or with parties also having lodged reservations. This possibility is highly relevant for the listing of marine mammal species, as we will see.

All whale species (*Cetacea spp.*) are listed on the Appendices of CITES. While general trade limitations are in place for the genus itself due to its listing on Appendix II, several whale species are also listed on Appendix I, which means that all international trade is prohibited: bowhead and right whale; all rorquals (except for the West Greenland stock of minke whales (*Balaenoptera acutorostrata*) which is listed on Appendix II); sperm whale; and gray whale. In other words, pursuant to the moratorium on commercial whaling under the IWC, which applies to the 'great whales', these species are also included in Appendix I of CITES. Moreover, several dolphin species, one river dolphin species, three porpoises, two families of beaked whale and bottle-nose whale, as well as the pygmy right whale (*Caperea marginata*) are listed. Not surprisingly, especially the whaling nations Japan, Iceland and Norway, and in some cases the Pacific micro-state Palau, have lodged several reservations against the listing of these species. Given that Norway and Iceland have also objected to the whaling moratorium they are free to commercially hunt and trade in whale products. Prior to leaving the IWC, Japan, on the other hand, was not able to commercially hunt whales and place them on the international market, but given its reservations to the CITES listings, it has been able to purchase whale meat from other states. It thus does not come as a surprise that commercially hunted whales in Iceland have been actively traded with Japan. Particularly fin whale meat that can be found in Japanese restaurants often stems from Iceland.

The situation concerning seals is different. Pinnipedia themselves are not listed on the Appendices. Instead, four species can be found on all three Appendices: monk seals (*Monachus spp.*) on Appendix I; fur seals (*Arctocephalus spp.*) on Appendix II, with the exception of the Guadelupe fur seal (*Arctocephalus townsendi*) which is listed on Appendix I; and the walrus (*Odobenus rosmarus*) which is listed on Appendix III based on Canada's request for international support. No reservations against the listing of these species exist since no country is actively pursuing these species for the purpose of international trade.

For the sake of completion, the dugong (*Dugong dugon*) and three manatee species are listed on Appendix I. Palau has lodged a reservation against the listing of the dugong.

The listing of the polar bear has caused much uproar within CITES. This dispute amongst the parties is a synonym for the difficulties that CITES faces. One of the main points of criticism is that a two-third majority decides over the future of international trade in a specific species. Even though the range state may be firmly opposed to the decision—either for commercial reasons or for reasons taking into account the interests of the local populations—it may be outvoted by states that have themselves no immediate interest in or benefit from such trade. I will delve more into the listing of the polar bear in Sect. 7.7.

The setting of CITES meetings is oftentimes a highly emotional one. After all, CITES deals with species of high emotional value, such as elephants, polar bears or whales. Not surprisingly, CITES meetings are the battleground of international organisations opposing and supporting the trade in products from species such as the African elephants. Moreover, decisions that are taken at CITES also affect other regimes, such as the whaling regime, and vice versa. For instance, Japan's now discontinued research whaling of sei whales (*Balaenoptera borealis*), which are listed on Appendix I, and the associated sale of whale meat on the Japanese market was debated at the meeting of the CITES Standing Committee—a committee setting the overall course for the conferences of the parties—in October 2018. While in principle legal, and arguably even required under the whaling convention in order not to waste whale meat, the Committee found this to be in violation of CITES since this *de facto* constitutes commercial trade despite an Appendix I-listing. As a response, Japan promised to take action by delaying the departure of the vessels and to halt the research hunt for sei whales.[4] While Japan did indeed take remedial action, this was not reflected in media outlets against whaling. For instance, the *International Marine Mammal Project* stated that "CITES Rules Japan Whaling Violates International Law."[5] This is bold at best, but legally also wrong since it infers that all Japanese whaling is in violation of international law—which branch of international law remains in the dark. The CITES Standing Committee did find that research whaling of sei whales is in contravention of CITES, but it did not find that other whaling is in contravention of CITES or other regimes. As we will see in Chap. 5, scientific whaling is legal under the whaling convention while also the sale of whale meat deriving from these whale hunts is permitted.

This case demonstrates how complex legal issues are broken down into simple, if not incorrect, statements to influence public opinion. The electronic form that visitors of the story above can fill out in order to show their opposition to whaling is but one example in this context. Not surprisingly, CITES meetings commonly see intense lobbying efforts, the display of large posters, as well as other means to influence the voting behaviour of states. As could be expected, also the decision of the CITES parties not to uplist the polar bear from Appendix II to Appendix I was received with much resentment by humane and animal welfare organisations. In an interview, a high-level executive of the *International Fund for Animal Welfare* (IFAW), speaking on behalf of several other organisations, even noted in regard to the decision to leave the polar bear on Appendix II: "In years to come, people will

---

[4]CITES (2018).

[5]International Marine Mammal Project (2018).

look back on this moment with great shame."[6] This neglects the fact that the Polar Bear Specialist Group (PBSG) of the IUCN along with other science-based organisations did not support the proposed transfer.[7] In Chap. 8 we will delve more deeply into the role of non-governmental organisations in shaping and influencing marine mammal-related legal regimes.

In conclusion, CITES is a regime which plays an important role for the management of marine mammals, particularly whales and polar bears. While the management of international trade has a questionable impact on improving or maintaining the conservation status of a species, it nevertheless sets important precedents for public discourse. For instance, if CITES had after all decided to uplist the polar bear to Appendix I and had thus banned all trade in polar bear products, also other fora that deal either with climate change or species protection, along with fora dealing with human rights, would have been impacted by this decision. On the one hand, it might have strengthened the no-use public discourse, on the other it might have negatively affected CITES itself since the human rights dimension might have led to action from the Arctic's indigenous peoples to challenge this decision. The cases before the European Court of Justice to overturn the EU seal trade regime stand exemplary in this regard (see Sect. 6.7.2). Another pathway might have been to split-list the polar bear: only those sub-populations that are in fact declining could be transferred to Appendix I while the others remain on Appendix II. This did occur in the case of African elephants as well. However, these steps remain in the realm of speculation. For the time being, polar bears remain on Appendix II while the 'great whales' remain on Appendix I.

4.2.2 The Conservation of Migratory Species

The regulation of trade as a means to ensure a favourable conservation status of a species is a rather uncertain method. Already the Stockholm conference of 1972 therefore also called for other means. While CITES is to a large extend based on species occurring in a nation state's sovereign territory, many animal species don't consider national borders, but migrate across them or are even primarily at home in areas beyond national jurisdiction. These migratory species and species in international waters are those species that the Stockholm conference sought to protect.[8] In 1979 the Convention on Migratory Species of Wild Animals (CMS)[9] was adopted, directly responding to this request. The convention currently has 127 parties. The United States, Canada, China, Japan and the Russian Federation, amongst others, are not party to the convention. While the operational and functional scope of the CMS

[6]IFAW (2010).

[7]PBSG (2013).

[8]See United Nations (1972), Rec 32.

[9]Convention on the Conservation of Migratory Species of Wild Animals, 23 June 1979 (1651 UNTS 333).

is different to CITES, it furthermore includes an ecosystem dimension: focusing on specific species, it also aims to conserve the habitats in which they dwell. The ecosystem approach is therefore a fundamental principle that is embedded in the convention. Moreover, the principle of intergenerational equity—the passing on of the world to the next generation in either the same or a better state—is deeply enshrined in the convention. The Preamble notes that "each generation of man *holds* the resources of the earth for future generations and has an *obligation to ensure* that this legacy is conserved and, where utilized, is used wisely."[10] The present generation is therefore in control of the way resources are being used and is required, in a legally-binding fashion, that the following generation is able to use these resources equally.

We must remember that the CMS is a convention that was concluded at the end of the 1970s, more than 10 years before the Rio Summit of 1992 where environmental principles were codified in the Convention on Biological Diversity (CBD) and other treaties and declarations. The CMS is therefore a progressive instrument that has integrated means and ways of species conservation that previous regimes, such as CITES, have not.

In order for the CMS to work it functions in a similar manner as CITES and contains two Appendices on which certain species are listed. The species listed on Appendix I are those species which are endangered and which are to be spared from any takes. Four exceptions to this prohibition can be found: if the take is for scientific purposes; if the take benefits the propagation and thus the conservation of the species; if it is conducted by subsistence users; or if other extraordinary circumstances require a take. Moreover, states are required to protect and restore the habitats of these species, and prevent or remove any impediment to their migration. We see here that the ecosystem approach takes precedent. Interestingly, the CMS does not make exclusive reference to scientific evidence that proves the endangered status of the species. Instead, it refers to "reliable evidence, *including* the best scientific evidence available."[11] While this is not explicitly referred to, 'reliable evidence' could also refer to the traditional knowledge of indigenous and local populations which is, at least in modern times, often considered as reliable as scientific evidence.

Species listed on Appendix II are those migratory species which are not endangered but whose conservation status is unfavourable. In order to ensure their continuity, range states of these species are to enter into species-specific 'Agreements' and foster international cooperation for their protection. The fact that the CMS encourages the conclusion of separate Agreements to reach the convention's objective identifies it as a framework convention. This means that it provides for the wider framework of international cooperation but leaves it up to the parties to decide on the nature of more specific agreements. Moreover, these Agreements may also include non-parties to the CMS. In other words, even if, for

[10]Own emphasis.

[11]Own emphasis.

example, the US are not party to the convention itself, it may still be a member of one of the Agreements (which it in fact also is in the case of three MoUs). The CMS enables species to be listed in both Appendix I and II, meaning that while a species may be shielded from all takes, parties can enter into more specific agreements for its protection. In a way this corresponds to the split-listing of CITES: while one sub-population may be listed in Appendix I, another may be listed in Appendix II. While that may be the case, CITES has inserted this split-listing after long debates only in the late 1990s. The listing in both Appendices of the CMS has existed from the beginning, once again highlighting the progressive nature of the convention.

Not surprisingly, quite a large number of marine mammals are listed on the Appendices of the CMS. In fact, the 77 listed whale species constitute the large majority of all whale species in the world. Of these, 15 species are listed on Appendix I (of which 8 are listed in both Appendices). All great whales, except for the gray whale, that are managed under the auspices of the IWC are listed here. The blue, bowhead, right and humpback whales have been listed since 1979 while the sperm, fin and sei whale were added in 2002. Since Japan is not a party to either the convention itself nor any of the instruments, it is not legally bound by the provisions set forth. Iceland is equally not party to the CMS and merely a party to the Agreement on African-Eurasian Migratory Waterbirds. Iceland's whaling is thus not affected by the CMS. Norway, on the other hand, has been party to the CMS since 1985. But since Norway does not whale species that are listed in Appendix I, it is not affected by the prohibition on any takes either. Norway's whaling activities in the North Atlantic focus primarily on the common minke whale (*Balaenoptera acutorostrata*), which is protected under the Agreement on the Conservation of Cetaceans of the Black Sea, Mediterranean Sea and contiguous Atlantic area (ACCOBAMS) (see Sect. 5.4). Since Norway is neither a range state of this agreement and thus not a signatory, it is not affected by it.

The case of seals is somewhat less dramatic. All in all, only six seal species are listed on the Appendices. Only the Mediterranean monk seal (*Monachus monachus*) and the Caspian seal (*Pusa caspica*) are listed on Appendix I and II while the other species are listed on Appendix II. However, since neither is affected by ongoing large-scale seal hunts, direct taking of these species is not a major threat. The relevance of the CMS and its agreements thus lies in the alleviation of multiple threats stemming from climate change, infrastructure development or fishing. We will get to know the instrument for the protection of the Mediterranean monk seal in Sect. 6.4.4.

Lastly, and not surprisingly, the polar bear has been listed on Appendix II since 2014. Although, as we have seen above, once listed in Appendix II the parties to the CMS are to take specific steps for the conservation of the species, no such Agreement has been concluded in case of the polar bear.

In conclusion, although the CMS places great emphasis on cetaceans, the CMS as a regime has not had a lasting impact on global whaling activities. The reason is rather simple: the whaling nations are either not party to the convention or not party to any of its Agreements. The threat reduction to marine mammals can therefore be found in the CMS' role for ecosystem and habitat protection. This occurs predominantly through the Agreements that have been concluded under the convention, all of which we will get to know below. This is further underlined by the fact that a the 12th Conference of the Parties, held in 2017, the parties adopted a resolution endorsing the new concept of Important Marine Mammal Areas (IMMA). An IMMA is a concept developed by the IUCN's Marine Mammal Protected Areas Task Force and is defined as "discrete portions of habitat, important to marine mammal species, that have the potential to be delineated and managed for conservation."[12] This means that when setting out plans for the conservation and management of specific areas, special attention is paid to marine mammals in these areas. While the Task Force addresses national governments, this consideration for marine mammals is also to be undertaken by intergovernmental organisations, conservation groups and the public. By adopting this resolution, the CMS parties have therefore demonstrated that marine mammal protection is to occur independently from their listing on the Appendices and rather as a normative standard in conservation.

4.2.3 The Law of the Sea

The vastness of the oceans has since time immemorial triggered human imagination. The seemingly endless horizons and the equally seemingly endless resources have made humankind use the oceans from the coastal waters to the high seas. Quite naturally, conflicts on resource use and the right to access to resources has accompanied this use. A branch of international law, the law of the sea, is a remedy against these conflicts and is a means to govern the global commons. The law of the sea is therefore not a phenomenon of the twentieth century but has been accompanying the development of international law in general for several hundred years. Although Hugo Grotius (1583—1645) is generally considered the father of the modern version of international law, also other scholars and thinkers of the fifteenth, sixteenth and seventeenth centuries have markedly shaped modern international law by drawing on a mixture of reason, custom and scripture. But while these were elementary in shaping international law, it was after all Hugo Grotius whose work *Mare Liberum* from the year 1609 established customs and practices that are still relevant under the modern law of the sea. Most notably, the freedom of the seas, which Grotius established to substantiate a counter-claim to the Portuguese who aimed to maintain trade monopoly in the Far East, is still a guiding principle of the law of the sea today.

Starting in the seventeenth century, European governments started to keep records of diplomatic exchanges, thereby establishing precedents and a record of customs. However, only in 1873 when the International Law Association was established, the first attempts to codify the international law of the sea, up till then a fragmented imbroglio of scholarly perception, were undertaken. But only in the

[12]Marine Mammal Protected Areas Task Force (Undated).

twentieth century, the codified law of the sea saw its emergence in multilateral conferences: in 1924 and 1930, held by the League of Nations; and in 1958 and 1973 held by the United Nations. The 1973 conference culminated in the United Nations Convention on the Law of the Sea (UNCLOS),[13] which is also referred to as the 'constitution of the oceans' and which constitutes one of the most important international legal instruments of the modern age. While there were significant developments throughout the different conferences towards the contemporary regime, for instance the establishment of the 200 nautical mile exclusive economic zone (EEZ), the UNCLOS furthermore holds provisions that explicitly deal with marine mammals and even more explicitly with whales (cetaceans).

In order to understand the special status of marine mammals under the UNCLOS, let us take a brief look at the different maritime zones and the associated rights of the coastal and other states. Briefly summarised: the UNCLOS establishes several maritime zones, the most important of which for the purposes of this book I mention here: internal waters; the territorial sea; the exclusive economic zone; and the high seas. The coastal state has exclusive jurisdiction over the waters and the resources occurring in the internal waters (e.g. rivers, estuaries, bays or lakes) and other states need explicit permission from the coastal state to access these waters and to make use of the abundant resources. Similarly, the territorial sea, which reaches up to 12 nautical miles (c. 22.2 km) from the coastline into the sea, is under the exclusive jurisdiction of the coastal state. This means that in order for other states to make use of the resources, permission of the coastal state needs to be acquired as the coastal state enjoys full sovereignty over these waters. Contrary to internal waters, however, the principle of 'innocent passage' prevails in the territorial sea. Innocent passage is a form of the freedom of the seas and describes the passing of vessels through the territorial sea with non-adversarial intent and maintaining peace and order. The EEZ stretches 200 nautical miles (c. 370 km) from the baseline and provides the coastal states with sovereign rights over the subsea and subsoil resources. The surface waters, however, are free for other vessels to use. What makes the determination of the EEZ particularly difficult is the existence of physical continental shelves, which are also legally relevant. Under the UNCLOS, based on complex criteria and determination process, a coastal state has the possibility to justify the extension of its continental shelf up to 350 nautical miles (c. 648 km) from the coast. It is this possibility which has led to the assertion of conflict in the Arctic since several Arctic Ocean coastal states indeed justify their EEZs extending towards the North Pole.[14] Be that as it may, the high seas constitute a major maritime zone, which, as the name implies, is not associated to any coastal state. Unless there are specific agreements in place, for instance regarding fishing or marine conservation, the freedom of the seas prevails in the high seas, allowing all states to freely roam and use the oceans.

With this brief overview over the maritime zones of the UNCLOS, let us now turn towards the marine mammal provisions. Very broadly speaking and applicable for

[13]United Nations Convention on the Law of the Sea, 10 Dec 1982 (1833 UNTS 3).

[14]Byers (2013), pp. 107–109.

all marine living resources, the UNCLOS promotes the conservation and 'optimum utilisation' of these based on scientific findings. In principle, therefore, it appears that also marine mammals are considered under this clause. But life would not be life if it was that simple. Article 64 establishes special rules for highly migratory species. These species are listed in Annex I of the UNCLOS and include, next to several fish species, all cetaceans, including dolphins and monodontidae (beluga and narwhal). Interestingly, the Annex separately lists "Dolphin: Coryphaena hippurus; Coryphaena equiselis." This, however, does not refer to the marine mammal, but to the dolphin *fish*. Notwithstanding, the article requires states to cooperate in 'appropriate international organisations' to ensure the conservation and optimum utilisation of these species. If there are regions in which no such organisation exists, but which is subject to active harvest of these species, the states engaged in this harvest are required to establish such organisation. The Inter-American Tropical Tuna Commission, for example, is such an organisation (see Sect. 5.5.1).

But the requirements of coastal states to work together for the conservation (and arguably optimum utilisation) go further and are further stipulated in articles 65 and 120. Article 65 elaborates on marine mammals and the clause of optimum utilisation in article 64. Here, the role of optimum utilisation is diminished and the conservation of marine mammals required by the states as the goal of cooperation. After all, the article notes that the role of an 'appropriate international organisation' may also be to "prohibit, limit or regulate the exploitation of marine mammals." In relation to whales, states are required to work through these organisations for the "conservation, management and study" of whales. In other words, states are *not* required to cooperate to ensure that whales are not endangered, which would arguably also entail their sustainable use. The insertion of this article is, however, not surprising. We must remember that the UNCLOS was negotiated during the 1970s when the call for a whaling moratorium grew louder. Indeed, the UNCLOS was adopted in the same year the IWC adopted the moratorium on commercial whaling. To put the marine mammal provisions of the UNCLOS in easier words: when an international organisation, which is a requirement for marine mammal conservation, decides to prohibit whaling, parties to the UNCLOS have to obey this prohibition in spite of the UNCLOS' optimum utilisation clause. Article 120 even goes further: although the freedom of the seas applies in the high seas, this does not apply to the utilisation and conservation of marine mammals. After all, the article establishes that the provisions in article 65 are also applicable on the high seas.[15]

From the above we can see that although the UNCLOS has established certain rights concerning resource use within specific maritime zones, these do not apply to marine mammals and particularly not to whales. Although scientific findings are supposedly the fundament of management and utilisation, the UNCLOS requires states to cooperate on marine mammal issues to ensure their conservation. One might argue that Japan's withdrawal from the IWC in December 2018 breaks with the requirement to cooperate, if 'cooperation' means membership in an organisation.

[15]Churchill and Lowe (1999), pp. 317–320.

After all, Japan resumes commercial whaling in its own EEZ despite the existence of the whaling moratorium. Although the UNCLOS does not explicitly refer to the IWC as the only appropriate international organisation for whales, Japan is still required to cooperate further what some commentators have interpreted as meaning either joining another organisation or starting a new one. Otherwise, it would be in non-compliance with the requirements for marine mammals under the UNCLOS. At the moment, only one other organisation for the management and sustainable use of marine mammal exists: the North Atlantic Marine Mammal Commission (NAMMCO). Since NAMMCO only deals with marine mammals in the North Atlantic, a Japanese membership is not possible (see Sect. 4.3.2). This being said, the UNCLOS does not define the term 'cooperation' more closely. While not being a member of the IWC anymore, Japan has not broken with it completely and maintains observer status with the right to participate in meetings and to contribute to the discussions (yet not with a right to vote) while still working on scientific issues with the IWC's Scientific Committee. Furthermore, Japan is an observer to NAMMCO and regularly participates in its meetings. One might therefore argue that it *does* cooperate on marine mammal issues. Whether this interpretation is correct remains to be seen.

In conclusion, while the UNCLOS does foster the conservation and optimum utilisation of marine living resources, this does not apply to marine mammals, and particularly not to whales. The conservation of whales stands at the fore of the requirements enshrined in the UNCLOS and once the IWC (or any other appropriate international organisation) decides to prohibit whaling, states are required to comply.

4.3 Regional Regimes

4.3.1 Europe

When attending the meetings of the IWC, those Commission members that are also members of the European Union attract the observer's attention due to their strong view on the Commission being a conservation and not a sustainable use organisation. Also in side meetings, representatives of the European Union make unmistakably clear that any resumption of commercial whaling, or any whaling that could in any way be interpreted as commercial whaling, is not to be resumed. Anything else cannot be accepted.

The EU has put in place a regime that makes no exception to the credo of marine mammal protection. For instance, when Iceland applied for EU membership in 2009, which was ultimately voted down in an Icelandic referendum, one of the preconditions was that Iceland give up its whaling operations. Yet, under EU law there are differences between whales and seals. For the latter, it is the regime on trade in seal products, which started to emerge since 1983, that aims to end the trade in seal products in order to prevent seals from being killed for commercial purposes. We will get to know this regime in detail in Chap. 6. For now, suffice it to say that in

1983 the European Communities (the predecessor of the EU) banned the trade in products stemming from seal pups. In 2009, the EU banned the trade in seal products generally. For whales, several EU-wide regimes are in place that prohibit the deliberate disturbance (which includes the killing) of whales in EU waters and which prohibits the importation of whale products taken from the wild.

The basis for the EU's environmental policy can be found in the Treaty on the Functioning of the European Union (TFEU), one of the core treaties of European cooperation.[16] The TFEU requires EU member states to work together on global environmental problems, which includes the effective protection of species worldwide. To follow this doctrine, the EU is furthermore required to harmonise its rules concerning species protection. At the core of EU whale protection stands, therefore, the Habitats Directive, which was adopted in 1992.[17] The objective of the Habitats Directive is the contribution to biodiversity protection through both the conservation of natural habitats and through the protection of specific species. While the Directive aims at comprehensively managing specific areas, it furthermore lays down prohibitions for species listed in its Annex IV, including capture, killing and disturbance. Here, also the prohibition of importation of species taken from the wild is stipulated. All cetacean species are listed in this Annex. With this strict protective legal framework in place, it is not surprising that EU-members of the IWC apply a strict no-use approach to whales. However, we must bear in mind that the Directive does not provide for any criteria for the listing or de-listing of species onto the Annexes. Although the Directive recognises that scientific and technological knowledge is crucial for its implementation, it does not require science to be the basis for species listing. In other words, the listing of cetaceans can be purely political.

A second core regime is the regulation which implements CITES, Council Regulation 338/97/EC on the protection of species of wild fauna and flora by regulating trade therein.[18] Without the need to go further into detail it suffices to say that the regulation directly takes over the provisions of CITES, including its Appendices. Therefore, also whale species are banned from the EU's internal market. However, the regulation goes somewhat further than CITES: although CITES lists some whale species in Appendix I and II, the regulation lists all cetaceans in Annex A, which corresponds to Appendix I of CITES. By doing so, the EU has underlined its no-use approach to whales.

While the above refers to the European Union and its member states, the heading of this section is entitled 'Europe'. This means that there are also other regimes in place that are relevant not only for the European Union, but for countries that are located on the European continent, but that are not necessarily EU members.

---

[16]Consolidated version of the Treaty on the Functioning of the European Union, OJ C 326, 26.10.2012, pp. 47–390.

[17]Directive 92/43/EEC on the Conservation of Natural Habitats and of Wild fauna and Flora, OJ L 206, 22.7.1992, p. 7.

[18]Council Regulation 338/97/EC on the protection of species of wild fauna and flora by regulating trade therein, OJ L 61, 3.3.1997, p. 1.

Table 4.1 Difference between European Union and Council of Europe

The European Union	The Council of Europe
- A body consisting of 28 European states (at the time of writing, Brexit, the exit of Great Britain, has not been implemented)	- 47 member states, 28 of which are EU members
- Comprising decision-making institutions such as the European Parliament and European Commission, making binding decisions	- Primarily concerned with the promotion of human rights, democracy and rule of law
- Sharing a common market and customs union	- A forerunner, leading to European integration
- Guarantees 'four freedoms': free movement of goods, capital, services and labour	- Cannot make binding decisions other than treaties negotiated under its auspices

Especially the Council of Europe (CoE) is relevant for this book (see Table 4.1) since it was this body that negotiated and finally adopted the Convention on the Conservation of European Wildlife and Habitats (Bern Convention)[19] in 1979.

The Bern Convention is a treaty which was concluded as a means to foster cooperation on the protection of flora and fauna in Europe. Since particular emphasis is put on the protection of migratory animals, it is not only European countries that are party to this convention, but also several Northern African states. Also the European Union is party to the convention. The only CoE members that are not parties are the Russian Federation and San Marino.

While the responsibility to implement the convention lies with the nation states, a Standing Committee meets back to back with the meetings of the Council of Europe and issues recommendations on the best ways to implement the convention. The Standing Committee comprises members of each party with one vote per party.

Although the Bern Convention carries the term 'habitat' in its title, it is predominantly species-based. Three appendices list the species for which different types of protection are to be applied. Appendix I lists merely flora species and will be neglected here. Appendix II and III, respectively, differentiate between 'strictly protected' species and 'protected' species. For strictly protected species on Appendix II, all deliberate killing or capture, disturbance as well as trade is prohibited. Several marine mammal species are listed in this Appendix. Apart from the walrus, the monk seal and the Saimaa and Ladoga seal, all large whale species, the narwhal, the orca, pilot whales and several other dolphin species are listed. Moreover, all bear species, including the polar bear, are listed. The Bern Convention therefore implements both CITES and IWC provisions and places cetacean species under strict protection. All in all, 30 whale species are listed. We must note though that several whale species are not considered globally, but regionally. For instance, although the sperm whale is listed on Appendix II, it merely refers to the populations in the Mediterranean.

[19]Convention on the Conservation of European Wildlife and Habitats, 19 September 1979 (1284 UNTS 209).

Species listed on Appendix III are species that are to enjoy special conservation measures. Here, the parties are required to control the exploitation of these animals in a way that ensures a favourable conservation status. Closed seasons, temporary prohibitions and the regulation of trade are means by which this is to be achieved. Six seal species, including the harp and hooded seal, can be found on this appendix. Concerning whales, the Appendix is rather sweeping: "All species not mentioned in Appendix II." In other words, all whale species are in one way or another protected by the Bern Convention while also some seal species that are faring well are to be regulated. Indeed, as we have seen in Chap. 2, the harp seal is far from being an endangered species in any of its three subpopulations.

When considering the Bern Convention, two drawbacks in the analysis arise: first, I cannot ascertain when a specific species was put on any of the appendices, therefore not allowing me to place their listing in a wider context of marine mammal utilisation or conservation. Were whales listed from the very beginning or is this a more recent phenomenon? Second, I am not able to find any guidance on how the Bern Convention responds to calls for population culls, for example for seals. Arie Trouwborst and others have explored this issue in relation to the Norwegian wolf policy, which involves a winter wold hunt despite its strict protection under Appendix II of the Bern Convention. The authors have concluded that Norway is, at least in parts, at odds with the convention but that there are indeed several issues that remain unresolved.[20] Therefore, the seemingly clear approach to conservation that the Bern Convention applies, i.e. to take "requisite measures to maintain the population of wild flora and fauna at, or adapt it to, a level which corresponds in particular to ecological, scientific and cultural requirements," may be at odds with the strict protection provisions for species listed in Appendix II.

Be that as it may, from the above it has become clear that within the European Union as well as for the European continent the protection of marine mammals ranges high on the environmental agenda. While, potentially, the Bern convention might see amendments to the appendices, which would list the strict protection of some marine mammal species, other regimes are in place for the European Union that still affects importation, killing and trade. Either way, particularly concerning whaling, the EU has adopted a common position of all its IWC member states at IWC meetings.

4.3.2 The North Atlantic

The North Atlantic Ocean has seen a large number of international regimes dealing with the marine environment, with fishing and with marine mammals. As we will see in Chap. 6, it is especially seal hunts which have been regulated in the North Atlantic. This is hardly surprising. After all, the North Atlantic Ocean is the adjacent ocean to the European continent—a continent that is home to peoples having

[20]Trouwborst et al. (2017).

explored and colonised large parts of the world via international waterways. Since seals were and are an abundant North Atlantic species and subject to large-scale hunts, it is the North Atlantic which has seen the first multilateral agreement on sealing—the Jan Mayen Sealing Treaty in 1875 (see Sect. 6.2). Whales, on the other hand, have not been treated regionally until the conclusion of the CMS and its Agreements. Instead, whaling states rather opted for global instruments to deal with whales and whaling. Since polar bears and polar bear hunts did not and do not play a major role in the North Atlantic region other than in Svalbard and Greenland, it is only countries in the Arctic region that deal with polar bear issues (see Chap. 7).

Since this book deals with marine mammal regimes specifically, I will not delve deeply into regimes that consider marine mammals merely in passing. Meaning, although those regimes, for example fisheries regimes and regimes for the protection of the marine environment, are also relevant for marine mammals, they are not exclusively focusing on them since their existence is rooted in a different purpose. One of the most influential organisations in this regard is the International Council for the Exploration of the Sea (ICES). ICES is the oldest maritime organisation in the world and was already established 1902. While primarily concerned with scientific cooperation in the North Atlantic, ICES is an organisation which is open to all states adjacent to the Baltic and North Seas. Since the organisation is science-based, it does not have a regulatory purpose, but rather advises governments and other organisations regarding marine management. Inevitably, also marine mammals play a role and the ICES Working Group on Marine Mammal Ecology (WGMME) provides this advice on marine mammals. In this sense, the WGMME and ICES are not directly involved in the application of marine mammal relevant laws, but they are relevant for the shaping of rules. For instance, management decisions taken by NAMMCO also derive from the close cooperation between these organisations. Similarly, ICES works together with fisheries regimes, such as the North Atlantic Fisheries Organization (NAFO). Together with NAFO, ICES therefore influences domestic decisions regarding marine resource extraction as well as in collaboration with the members of the respective organisation. This was particularly relevant prior to the establishment of the EEZs in the 1970s when NAFO's predecessor, the International Commission for the Northwest Atlantic Fisheries (ICNAF), also set quotas concerning sealing in waters beyond the territorial sea. Here, scientific findings on seals as a species in the Northwest Atlantic marine ecosystem translated directly into management decisions.

While that may be so, the year 1992 was a crucial year for the management and the protection of marine mammals in the North Atlantic. In April 1992, the Agreement on Cooperation in Research, Conservation and Management of Marine Mammals in the North Atlantic (NAMMCO Agreement)[21] was adopted, establishing NAMMCO. The NAMMCO Agreement is rather unique in international environmental law since it was concluded between Iceland and Norway, as

[21] Agreement on Cooperation in Research, Conservation and Management of Marine Mammals in the North Atlantic, 9 Apr 1992.

well as Greenland and the Faroe Islands, both of which have autonomy within the Kingdom of Denmark, but are not independent states. Since Greenland and the Faroe Islands have the right to manage fisheries and other natural resources, they were able to enter into the agreement. This is contrary to the IWC where Denmark represents itself as well as Greenland and the Faroe Islands. After all, at the time of conclusion of the International Convention for the Regulation of Whaling in 1946, Greenland's and the Faroe's level of autonomy had not been established.

Be that as it may, the four countries came together to cooperate on the conservation, rational management and study of marine mammals. NAMMCO's decisions are all taken by consensus, top-down decisions, which would potentially lead to a stall of the organisation, are being prevented. Additionally, NAMMCO members are all engaged in the utilisation of marine mammals. To this end, NAMMCO has become a spearhead organisation for the rational use of marine mammals based on 'hard' scientific facts. Indeed, science is one of the major outputs of the organisation, resulting in a large number of scientific studies on marine mammals in the North Atlantic. This includes population censuses, studies on by-catch, observation and inspection issues, as well as hunting methods and humane killing. While the work of NAMMCO is often species-focused, for instance the statuses and threats of individual species, it also deals with diseases such as seal worms. When NAMMCO was established, some commentators feared that it would undermine the legitimacy of the IWC. This, however, has not been the case: first of all, NAMMCO's membership amounts merely to four countries, while the IWC's membership is currently 89 states. Second, NAMMCO's sphere of influence is limited to the North Atlantic. Third, NAMMCO's mandate includes *all* marine mammals species in the NAMMCO area and not just whales. Therefore, instead of working against each other, both the IWC and NAMMCO cooperate closely. An interesting feature of NAMMCO is the close cooperation with marine mammal hunting communities and their right to sustainable marine mammal utilisation. While, as we will see below, the IWC provides for aboriginal exemptions, NAMMCO considers all marine mammal hunting communities legitimate.

The decisions taken by the NAMMCO Management Committee have direct impact on marine mammal management in the NAMMCO area. While NAMMCO itself is not the management body but proposes management recommendations to its member states, these recommendations are taken seriously. We must highlight once again that the decisions are all taken by consensus. Management of marine mammals in the NAMMCO areas therefore does not lie in the hands of NAMMCO but in the hands of the respective member state. However, NAMMCO's influence on the national management strategies regarding specific species and how they can be used is significant.

But as we will see, species-specific management is not the only tool by which marine mammals are protected: marine protected areas (MPAs) are a key strategy which is being applied to place marine mammals and their ecosystems under special protection—an issue which we will further explore in Sect. 4.4. A key regime for the North Atlantic, for which MPAs play an important role, is the Convention for the Protection of the Marine Environment of the North-East Atlantic (OSPAR

Convention), which was adopted in 1992.[22] The OSPAR Convention is not a regime that was concluded specifically with marine mammals in mind, but is a merger convention between the Convention for the Prevention of Marine Pollution by Dumping from Ships and Aircraft (Oslo Convention)[23] and the Convention for the Prevention of Marine Pollution from Land-Based Sources (Paris Convention)[24]—hence the merger-name between Oslo and Paris: OSPAR. In other words, the OSPAR Convention is a regime identifying pollution in its various forms as one of the main threats to the North Atlantic ecosystem. Here, the role of the OSPAR Commission, established under the convention, rests in the assessment of the marine ecosystem and associated conservation recommendations to the contracting governments.

For instance, the year 2013 saw three recommendations on further measures for the protection and conservation of northern right whales, bowhead whales and North Atlantic blue whales. While the recommendations were merely recommendations to the contracting parties and therefore not legally binding, they nevertheless have significant normative power. After all, the recommendations contain precise steps for further monitoring of the species as well as a timetable and a template for reporting and implementation of the recommendations. The European Union, along with several separate governments of EU member states, as well as Switzerland, Iceland, Luxembourg and Norway are party to the convention. Actions taken in the OSPAR area by the member states consequently have an impact on the North Atlantic ecosystem and marine mammal species. This is especially true in OSPAR's work on underwater noise, one of the main issues impacting whale behaviour, having led to mass strandings. All in all, OSPAR's work areas, including biodiversity, hazardous substances, human activities, offshore industries, radioactive substances as well as cross-cutting issues, and associated decisions and recommendations strongly affect marine mammals in the OSPAR convention area. The most direct effect, however, lies in OSPAR's marine protected areas, which directly impact any human activity therein. We will explore these in Sect. 4.4.

4.3.3 The Antarctic

By many the Antarctic is considered to be the last frontier on Earth. After all, Antarctica is a vast, isolated, ice-covered continent which is, apart from scientists and research personnel, uninhabited. Yet, from the outset, the Antarctic has been a 'goldmine' for marine mammal exploitation. Sub-Antarctic islands such as South

[22]Convention for the Protection of the Marine Environment of the Northeast Atlantic, 22 Sept 1992 (2354 UNTS 67).

[23]Convention for the Prevention of Marine Pollution by Dumping from Ships and Aircraft (Oslo Convention),15 Feb 1972 (932 UNTS 3).

[24]Convention for the Prevention of Marine Pollution from Land-Based Sources (Paris Convention), 4 June 1975 (1546 UNTS 119).

Georgia or Macquarie Island saw large-scale whale and seal hunts while also the Southern Ocean itself was the hunting ground for large whales. Until recently, Japanese research whaling targeted minke and fin whales in the Antarctic, which led to court proceedings before the International Court of Justice (ICJ) (see Sect. 5.3.5).

Contrary to other regions of the world, the Antarctic is governed by a complex array of different legal instruments. These instruments taken together are referred to as the Antarctic Treaty System (ATS). The ATS essentially comprises three treaties: the Antarctic Treaty (1959)[25] and its Protocol on Environmental Protection to the Antarctic Treaty (1991) (Madrid Protocol)[26]; the Convention for the Conservation of Antarctic Seals (1972) (CCAS)[27]; and the Convention on the Conservation of Antarctic Marine Living Resources (1980) (CAMLR).[28] Additionally, the recommendations adopted by the Antarctic Treaty Consultative Meeting feed into the ATS. Inevitably, marine mammals play a role in theATS. While these treaties are lumped together under the term 'Antarctic Treaty System', it is noteworthy that these are all independent treaties with their own independent membership. This means that countries can be party to one treaty, but not a party to another. New Zealand, for example, has not signed the CCAS, but is nevertheless one of the forerunners in Antarctic governance.

The ATS does not address marine mammals as a singular group of species, as the existence of the CCAS proves. The reason for the different treatment of whales and seals stems from the different histories of whaling and sealing in the Antarctic as well as the different international legal regimes that regulate these activities. Antarctic whaling has been conducted for a long time and international attempts to regulate whaling, i.e. the International Convention for the Regulation of Whaling, were in place when the Antarctic Treaty was adopted. Antarctic sealing, on the other hand, saw a sharp rise in the nineteenth century with drastic reductions of seal herds on the sub-Antarctic islands, but the industry never took off in a long-lasting manner. Also the international legal framework for sealing had not converged in one international agreement (see Chap. 6). Therefore, the treatment of whales/whaling and seals/sealing within the context of the ATS is different.

As regards whaling, the Antarctic Treaty itself is silent on any regulatory regimes or organisations. We must remember that at the time of the treaty's conclusion, whaling was internationally still very much accepted and although the IWC had put in place provisions to limit—or at least regulate—whaling internationally, the Southern Ocean was the area in which all whaling nations participated in the so-called 'Whaling Olympics' in order to reach maximum yields (see Sect. 5.3.3). Only when the tides turned in the 1960s and 1970s, also the different members of the

---

[25] Antarctic Treaty, 1 Dec 1959 (402 UNTS 71).

[26] Protocol on Environmental Protection to the Antarctic Treaty, 4 Oct 1991 (30 ILM 1455).

[27] Convention for the Conservation of Antarctic Seals, 1 June 1972 (11 ILM 251).

[28] Convention on the Conservation of Antarctic Marine Living Resources, 20 May 1980 (1329 UNTS 48).

Antarctic Treaty System started to respond. The result of the overall environmental awakening also in the Antarctic was the conclusion of the CAMLR and its inherent ecosystem approach to environmental protection. In other words, CAMLR aims to treat the Antarctic ecosystem comprehensively without specific focus on individual species. What is noteworthy, however, is that CAMLR explicitly singles out the whaling convention and the CCAS. This is to say that CAMLR holds provisions in which the rights and obligations of CAMLR parties that are also parties to the whaling convention and CCAS are to be respected. This means that CAMLR does not overrule the IWC and the decisions taken by CCAS members.

In a similar manner, the Madrid Protocol reinforces this approach. It appears therefore that both the CAMLR and the Madrid Protocol do not consider marine mammal issues. Particularly in the case of the latter, the prevalence of the UNCLOS is underlined in the context of marine resource extraction while other marine activities that the Protocol relates to deal with scientific research programs, tourism and other governmental and non-governmental activities. While that may be so, some commentators have suggested that the Madrid Protocol, and arguably also the CAMLR, can be applied to issues relevant for, but not directly related to, whaling activities. For instance issues of general environmental protection or cooperation in the planning of activities in the Antarctic Treaty area would fall under this category. However, one particular trademark of Antarctic cooperation is that decisions both under the Antarctic Treaty and under CAMLR are taken by consensus. Whaling is an activity which prevents consensus decision-making given its emotional nature. The IWC is the manifestation of failed decision-making. Therefore, ATS parties have by and large abstained from taking into consideration whaling issues in order to avoid a stall in negotiations and decision-making.

The importance of the ATS *vis-à-vis* Antarctic whaling became relevant when in February 2007 a fire on the Japanese whaling vessel *Nisshin Maru* killed one crew member and disabled the vessel for 10 days. This meant that it drifted without control through Antarctic waters, spurring the danger of major oil spills. New Zealand raised concerns about these potential environmental impacts and wished to discuss the issue amongst the Antarctic Treaty parties. Japan, however, quickly blocked these attempts, arguing that this would inevitably lead to normative discussions about whaling in Antarctica—an issue that would affect the cooperative spirit of Antarctic governance.

The case of sealing is somewhat different, as indicated in the opening paragraphs of this section. Since I enter into a more detailed discussion on Antarctic sealing in Sect. 6.4.3, I merely briefly introduce the regulatory regime here. In 1972 the CCAS was adopted, which put in place regulatory measures for the hunt of six Antarctic seal species, three of which were fully protected. Yet, since sealing at that time was no longer actively conducted in the Antarctic, the convention can be considered a proactive, or precautionary rather than a reactive measure. Also up to the time of writing, no seal hunts are ongoing in the Antarctic. However, also seals are considered differently than other species in the Antarctic, as the CAMLR and the Madrid Protocol show. While in the Antarctic sealing has not reached as emotional a level as in, say, Canada, the cooperative spirit of the Antarctic Treaty may also play a role while CCAS and sealing have not been addressed amongst the parties to the Antarctic Treaty.

Generally, although the ATS, legally speaking, does indeed provide for the inclusion of marine mammals into its regulatory regimes, the plain fact that marine mammal hunts have caused such great controversy amongst parties particularly in the IWC has prevented the ATS to engage with a full-scale ecosystem approach to Antarctic environmental protection. Marine mammals are singled out. While this might, on the surface, seem unreasonable, it makes quite a lot of sense. If sealing or, particularly, whaling were to enter Antarctic decision-making discourse, the consensus-driven regime would be in jeopardy. It appears therefore reasonable, if not necessary, to leave out such controversial issues to ensure the longevity and effectiveness of the regime itself.

4.4 The Importance of Marine Protected Areas

While in the Antarctic sealing and whaling have been largely avoided by the parties to the Antarctic Treaty System, this does not mean that whaling regimes have avoided the Antarctic. A crucial role in this regard play marine protected areas (MPAs), which have emerged as a conservation tool not only in the Antarctic, but all over the world. Currently, around 3.4% of the world's oceans are MPAs, mostly, however, in national waters. MPAs constitute the oldest type of mechanism for the protection of the marine environment. As the name implies, marine protected areas, in broad terms, are specifically designated marine areas, including coastal, estuary or lake areas, in which human activity is limited in regard to the exploitation and utilisation of the marine environment. This means that resource exploitation is either fully prohibited or confined to specific seasons or species. The objectives of MPAs are therefore to protect the marine ecosystem or to protect specific species. The terminology of MPAs varies greatly, however, and different countries use different terms, such as marine park, marine protected area, marine reserve, sanctuary or something similar.

Despite the existence of diverging terminology, the quintessence of these areas is the same and they denote areas in which in some form or another the marine ecosystem is protected. One special feature of MPAs is that they are always concluded under one specific regime—either multilateral or domestic. For seals, for example, Finland has put in place several seal sanctuaries in which seals are strictly protected. These sanctuaries are subject to national legislation. MPAs also play a role under the International Convention for the Regulation of Whaling and its Schedule (see in detail Chap. 5). Although the ICRW in principle allows for the commercial hunt of whales, since 1982 a zero catch quota has been put in place—commonly referred to as the 'whaling moratorium'. Apart from this quota, which can theoretically be lifted again, a second means of broadly protecting whales has been put in place: whale sanctuaries. Whale sanctuaries are specifically designated extremely large areas all over the world in which whales are fully protected irrespective of whether the moratorium is still in place or not. They therefore function as independent management schemes under the convention.

Whale sanctuaries have entered the discourse on whales and whaling already in the 1970s. In 1979, the first sanctuary was established in the Indian Ocean, encompassing the area of the entire Indian Ocean down to 55° South latitude. The second was the Southern Ocean Whale Sanctuary (SOWS) in 1995. The SOWS encompasses all waters surrounding the Antarctic continent up to 40° South latitude, but in the Indian Ocean up to 55° South latitude, thus joining with the Indian Ocean sanctuary. In these areas, all commercial whaling is prohibited. From the outset, Japan has opposed the establishment of whale sanctuaries, arguing that the concept would contravene the purpose of the whaling convention particularly since decisions are to be based on scientific findings. According to Japan, whale sanctuaries do not rely on science, otherwise they would not impose blanket protection for all whale species in a given area. Others, however, argue that whale sanctuaries are necessary bearing in mind the precautionary approach. Every 10 years, the IWC reviews the status of the whale sanctuaries, thus including the possibility to lift them. Up to the point of writing, however, this has not occurred. A third whale sanctuary for the South Atlantic Ocean has been brought on the IWC's agenda since the mid-1990s by several South American, and more recently, additionally some Western African states. The countries argue that potential whaling in the South Atlantic would significantly impact the wellbeing of their coastal communities while a whale sanctuary would have a beneficial effect, particularly with regard to whale watching and other tourism activities. Up to the point of writing, the proposal has failed to reach the necessary three-quarter majority within the IWC since this would mean a changing of the Schedule. Also a proposal for the establishment of a South Pacific Whale Sanctuary was rejected by the Commission in 2000. In this case, however, it is noteworthy that more than half of the Pacific Islands region is *de facto* a whale sanctuary since the EEZs of the countries with their countless smaller islands are so large and are nationally declared whale sanctuaries.

What needs to be remembered however, is that the national and IWC-declared whale sanctuaries are not considered MPAs in the sense of the IUCN. This is to say that the boundaries of these are artificially set while focusing merely on one specific group of animals. The IUCN approaches MPAs through an ecosystem-based lens, meaning that in-depth knowledge of the ecosystem, including its cultural role, are to be considered. National and IWC MPAs are therefore more of a political than of a conservation nature.[29] In this sense, the Japanese claim of being unscientific does not appear to be unreasonable. This is supported by the fact that the whale sanctuaries only relate to the lethal take, that is the hunt, for whales. They aim to put a halt to whaling. What they do not do is take into account other threats, such as food shortage, pollution or habitat degradation. In order to further advance the idea of a comprehensive conservation system, the parties to the Convention on Biological

[29]Hoyt (2011), p. 24.

Diversity (CBD)[30] have developed the concept of ecologically or biologically significant marine areas (EBSA) in the first decade of the twenty-first century. EBSA comprise the following criteria: uniqueness or rarity; Special importance for life history stages of species; Importance for threatened, endangered or declining species and/or habitats; Vulnerability, fragility, sensitivity, or slow recovery; Biological productivity; Biological diversity; and naturalness. Contrary to the concept of sanctuaries, EBSA are areas that are individually managed by cooperating states, following the rules of international law and in particularly the UNCLOS. In how far the implementation of EBSA will yield positive long-term results remains to be seen given the recent (and still ongoing) development of the concept.

While the approach taken by the CBD is global and therefore expectedly taking rather long to be fully implemented, other regimes have taken their own steps to put in place MPAs that take a comprehensive, ecosystem-based approach to conservation. One such regime is the OSPAR Convention. Here, MPAs play a crucial role for the overall success of the convention. OSPAR defines MPAs as "areas for which protective, conservation, restorative or precautionary measures have been instituted for the purpose of protecting and conserving species, habitats, ecosystems or ecological processes of the marine environment."[31] This includes waters within the EEZs of OSPAR parties as well as beyond the EEZs. Since 2010 seven MPAs in the high seas have been established, which are either co-managed by OSPAR parties or for which management plans are currently under development. Currently, all in all more than 400 MPAs have been established under OSPAR, covering around 6% of the convention area. The aim of the convention is to establish a network of MPAs that would ultimately result in the comprehensive management of MPAs with the ultimate goal of effective biodiversity protection. As we can see, OSPAR's approach is significantly different to that by the IWC. This is hardly surprising, however, given the IWC's limited mandate and OSPAR's broad, ecosystem-based mandate.

The CBD and OSPAR serve as but two examples for an MPA-based approach to the protection of the marine environment and numerous regimes, institutions and organisations promote the establishment of MPAs. An authoritative source in this regard is Erich Hoyt's *Marine Protected Areas for Whales, Dolphins and Porpoises.*[32] With the examples above, I think it has become clear that MPAs in their various forms play an important role for the protection of marine mammals as part of the protection of the marine ecosystem. With a sole focus on specific species, rather little can be achieved, particularly when it refers to the lethal take only. Therefore, comprehensive management is key, including pragmatic approaches that take into account not merely the protection of the marine environment, but also the human use of this environment. Marine spatial planning and comprehensive

[30]Convention on Biological Diversity, 5 June 1992 (1760 UNTS 79).

[31]OSPAR (Undated).

[32]Hoyt (2011).

ocean zoning, which accommodates human use and habitat protection are important tools in this regard.

4.5 Summary and Conclusion

The cooperative landscape on environmental issues has yielded a wide array of multilateral agreements that take two distinct approaches: the ecosystem-based *vis-à-vis* the species-based approach. In this chapter we have gotten to know regimes that are primarily species-based and often in relation to, human exploitation of these. This is particularly the case for whales.

CITES is probably the most prevalent example in this regard. Irrespective of the conservation status of the different species, all 'great whales' are listed under Appendix I, even though some populations are indeed stable. Paired with the special treatment of whales under the UNCLOS, which also limits the freedom of the high seas in regard to whaling, it becomes clear that on a global scale whales occupy a special role in the legal landscape. This is further underlined by the European (Union's) legal perception of whales. Once again, irrespective of the conservation status of whales, both the European Union and also most of the members of the Council of Europe protect cetaceans beyond considerations of science or human use. It seems, therefore, that the protection of whales which excludes human use (also referred to as 'preservation'), has become a normative standard in international law-making circles. At least it has become a standard amongst those states that markedly shape international environmental regimes.

Internationally, seals are somewhat a different story. Here, blanket protection, despite them being marine mammals, appears not to be as strict as is the case with whales. More attention is paid to conservation statuses and generally a more limited percentage of seal species is protected *vis-à-vis* whales. In the European Union, however, also seals are perceived beyond notions of sustainable use, as we will see in Chap. 6.

While there are differences between the regimes presented here, both in their way of functioning and in their way to approach marine mammals, it becomes clear that especially in global regimes, first and foremost CITES and UNCLOS, marine mammals play a special role. Since these regimes were drafted at a time when the opposition to marine mammal hunts was highly prominent, this does not seem surprising. Whether or not their timeliness in the twenty-first century in this regard is still guaranteed is certainly subject to debate.

Questions and Research Tasks

- Would the establishment of a South Atlantic Whale Sanctuary be beneficial for long-term whale conservation?
- Do marine mammals play a special role in the legal system of your country or region? If so, why?

- Are there any other groups of animals that enjoy such a high degree of internationally protection, irrespective of conservation status?
- Should there be a global regime treating marine mammals?

References

Byers M (2013) International law and the Arctic. Cambridge University Press, Cambridge

Churchill RR, Lowe AV (1999) The law of the sea, 3rd edn. Manchester University Press, Manchester

CITES (2018) CITES annual meeting sees States stepping up action against the over exploitation of precious timber and other species, Press Release 5 October 2018. https://www.cites.org/eng/news/CITES-annual-meeting-sees-States-stepping-up-action-against-the-over-exploitation-of-precious-timber-and-other-species_05102018. Accessed 23 May 2019

Couzens E (2013) Whales and elephants in international conservation law: a comparative study. Routledge, Abingdon

Hoyt E (2011) Marine protected areas for whales, dolphins and porpoises, 2nd edn. Earthscan/Routledge, Abingdon

IFAW (2010) Death Knell for Polar Bears as CITES Votes Against Conservation. https://www.prnewswire.com/news-releases/death-knell-for-polar-bears-as-cites-votes-against-conservation-88379517.html. Accessed 23 May 2019

International Marine Mammal Project (8 October 2018) CITES Rules Japan Whaling Violates International Law. http://savedolphins.eii.org/news/entry/cites-rules-japan-whaling-violates-international-law. Accessed 23 May 2019

Marine Mammal Protected Areas Task Force (Undated) Important Marine Mammal Areas (IMMAS). https://www.marinemammalhabitat.org. Accessed 23 May 2019

Orwell G (1945/1951) Animal Farm. Penguin Books, London

OSPAR (Undated) Marine Protected Areas. https://www.ospar.org/work-areas/bdc/marine-protected-areas. Accessed 5 June 2019

PBSG (2013) PBSG statement on proposed transfer of polar bear to CITES Appendix I. http://pbsg.npolar.no/en/news/archive/2013/CITES-PBSG-2013.html. Accessed 23 May 2019

Trouwborst A, Fleurke FM, Linnell JDC (2017) Norway's wolf policy and the Bern Convention on European wildlife: avoiding the "manifestly absurd". J Int Wild Law Policy 20(2):155–167

United Nations (1972) Report of the United Nations Conference on the Human Environment, Stockholm, June 1972. B. Recommendations for Action at the International Level. United Nations, Geneva. http://www.un-documents.net/aphe-b2.htm. Accessed 20 Apr 2017

Further Reading

Bankes N (2013) The conservation and utilization of marine mammals in the Arctic Region. In: Molenaar EJ, Oude Elferink AG, Rothwell DR (eds) The law of the sea and the polar regions. Interactions between global and regional regimes. Brill Nijhoff, Leiden, pp 298—321

Molenaar EJ (2003) Marine mammals: the role of ethics and ecosystem considerations. J Int Wildl Law Policy 6:31–51

Mossop J (2013) Marine mammals in the Antarctic treaty system. In: Molenaar EJ, Oude Elferink AG, Rothwell DR (eds) The law of the sea and the polar regions. Interactions between global and regional regimes. Brill Nijoff, Leiden, pp 267–292

The International Legal Framework for Whales

5

5.1 Introduction

Whales, or cetaceans, are probably some of the most iconic species in international discourse, having provided countless generations worldwide with awe and wonder. Not surprisingly, oftentimes 'the whale' is considered as being protected under international law, meaning that any sort of whaling is considered to be illegal. The International Whaling Commission (IWC) is thus perceived as being the watchdog over this alleged prohibition on whaling.

Unfortunately, reality is somewhat more complex. As a general rule we must remember that a treaty is only binding upon those states that have ratified it (*pacta tertiis*), meaning that it has been transposed into national law. States that have merely signed a treaty, are not supposed to act contrary to the treaty's purpose, but cannot be held accountable if they do. States that have not even signed a treaty are in no way bound by its provisions. This means that the IWC as a regulatory body only regulates those whaling activities of member states that are party to the International Convention for the Regulation of Whaling (ICRW) from 1946. But this convention we will get to know a little later in this chapter.

5.2 The History of International Whaling Regulations

For centuries whaling countries such as Iceland, Japan or Norway have regulated whaling activities in their respective waters. Particularly fisheries regulations stood at the fore of these regulatory instruments—after all, whales were considered fish as, for instance, the term 'whalefish', *hvalfisk* (Danish), or *Walfisch* (German) indicate. Be that as it may, we will not consider the national regulations for the whale 'fishery', but rather look at the international development of whaling regimes, which have gone parallel to the international discourse on whales. This is to say that ever since whales were discovered as a resource, Basque, Dutch and other whalers conducted intensive hunts for them, decimating their numbers drastically.

N. Sellheim, *International Marine Mammal Law*,
https://doi.org/10.1007/978-3-030-35268-4_5

Even though whaling activities had become significantly less profitable at the turn of the nineteenth to twentieth century, the link between overhunting and population decline was only inertly established—whether knowingly or unknowingly is impossible to determine.

Either way, with the decline in populations, whaling in the northern hemisphere also declined, and the main focus of international whaling activities rested on Antarctic pelagic whaling. Up to this point in the early twentieth century, no international, bi- or multilateral agreement that regulated the hunt for whales could be found—contrary to the hunt for seals, which we will explore in Chap. 6. Yet, by the 1920s it became clear that Antarctic whaling operations did not meet the expectations and that the sustainability of whaling was in jeopardy. On the one hand, whalers faced increasing difficulties in finding whales that would make their whaling journeys profitable. On the other hand, the global economic downturn made the sale of whale oil increasingly difficult. Even though national measures were in place also in the Antarctic—e.g. in the sub-Antarctic island of South Georgia which at that time was under British hegemony, specific whale species were protected, opening and closing seasons regulated—it became clear that international action was required to bring whaling back on its feet. It must be stressed that the focus at that time was not so much on restoring whale populations for the sake of conservation, but rather on restoring whale populations for the sake of revitalising the whaling industry.

5.2.1 The 1931 Convention for the Regulation of Whaling

Particularly two nations, who also quarrelled about the hegemony at sub-Antarctic South Georgia, had emerged as leading whaling nations internationally: Norway and the United Kingdom. Both countries had whaling regulations in place, but their approach to an international regime differed greatly. Norway's expertise in crafting legal instruments concerning whaling made it a forerunner in the development of international whaling regimes. The UK, on the other hand, believed that whaling is best regulated through individual bi- or multilateral agreements and not through a global regime. At the fore of the UK's argument stood its science-based approach which did not consider a global whaling regime on par with science-based decision-making. Since the 'camps' amongst the whaling countries could not agree on the best way to proceed in this matter, the issue was transferred to the Economic Committee of the League of Nations, which was the forerunner of the UN. Already in 1925 the League had initiated the process of determining a way forward as regards the use of marine fauna, including whales, and several conferences were held to consider global action and to discuss ways and means to ensure sustainable industries. In the end, at a 1929 conference, the League of Nations recommended the convention of an expert committee in Berlin in April 1930. Here, the first attempt to draw up an international whaling treaty was made, for which the Norwegian Whaling Act of 1929 served as a blueprint. 1930 furthermore marked the year in which the International Council for the Exploration of the Sea (ICES) established the Bureau of

International Whaling Statistics in Sandefjord, Norway. Upon voluntary reporting of whale catchers, the Bureau was able to collect data on the numbers, sizes, species, and areas of whales that were taken. In how far the reporting was accurate and complete cannot be ascertained.

The drawn-up text was not accepted as such, but circulated amongst the League's member and non-member states for commenting. Based on these comments, another Committee of Experts revised the first draft and presented its final report to the Assembly of the League of Nations. In this report it was made clear that the protection of whale stocks indeed served the interests of the whaling industry. With this overarching narrative guiding the debates surrounding the draft, the Convention for the Regulation of Whaling[1] was opened for signature on 24 September 1931 in Geneva. It was originally signed by 26 states, but was eventually ratified by 28. Even though today Japan is a major whaling nation, at that time its industrial scale whaling activities were in their infancy, which prompted Japan not to join the convention in order to further develop its whaling industry.

Since Norway and the UK were elementary players in the global whaling industry, the convention entered into force only upon ratification of eight states, which must have included Norway and the UK. Once the convention had been opened for signature, the UK, however, showed a degree of inertia in ratifying it since it feared that its, at that time, strict provisions may have the opposite effect of what it has set out to do: it may encourage whalers to change to so-called 'flags of convenience'—meaning they would sail under flags of nations that are not party to the convention—in order not to be bound by the convention's provisions. Whether or not this was a legitimate concern or whether other factors may have played a role in the UK's inertia is impossible to determine. Be that as it may, also the UK ratified the Convention on 18 October 1934 and it thus entered into force on 16 January 1935.

The Convention for the Regulation of Whaling was a convention that was drafted with the interests of the whaling industry at its core and thus with the protection of whales merely for this purpose. In other words, the convention did not strive to protect whales (or even ecosystems), but considered whales a resource that can be exploited for the benefit of humankind. This deeply anthropocentric approach was not unique for this convention, however, but can be found in many other bi- or multilateral regimes all over the world. Yet, particularly in the context of marine mammals, anthropocentric views shaped the regimes of that time, as we will also see in the case of the Bering Sea Fur Seal Regime (Chap. 6). Against this backdrop, the provisions of the convention must be read. Article 4, for example, puts the right whale and its subspecies under total protection. While seemingly beneficial for the species, we must remember that the term 'right' whale stems from the fact that it floats after being killed. It is thus the 'right' whale to go after, having led to exploitation that has driven this species close to extinction. The protection of the right whale is thus to be understood only for the possibility to exploit it further in the

[1]Convention for the Regulation of Whaling of 24 September 1931 (155 LNTS 349).

future. In the same manner, provisions protecting suckling whales, immature whales or female whales which are accompanied by calves must be considered.

The longevity of the whaling industry is furthermore at the heart of provisions of a more administrative character. For example, provisions for proper renumeration of whaling crews or the notification of national governments before engaging in whaling identify human and whale wellbeing merely as a 'side product.'

This being said, one might argue that the convention was rather progressive at that time, particularly with regard to the considerations of the local, indigenous populations. While the convention did establish certain regulations for the conduct of whaling, these did not apply to the 'aborigines dwelling on the coasts of the territories of the High Contracting Parties.' This meant that indigenous populations that hunted whales within the national boundaries of the parties to the convention were not subject to the restrictions and rules outlined in the convention. This was, however, only the case if they used only traditional boating propelled by oars or sail; if they did not carry firearms; if they were not employed by persons other than indigenous persons; and if they were not contractually bound to deliver whale products to third persons. If any of these conditions did not apply, they would have to be bound by the provisions of the convention.

Here we already see a manifestation of a discourse that has lasted throughout the twentieth and well into the twenty-first century: the dichotomy of commercial and subsistence use of marine mammals. The convention did not aim to regulate whaling as such, but rather industrial or commercial whaling—meaning whaling for profit. Since indeed profit was the guiding principle of the whaling industry at that time, this meant that the more whales could be killed and used, the more profit could be made, proving disastrous to whale populations and ultimately for the commercial whaling industry. Consequently, some kind of regulations were necessary to ensure the sustainability of whaling as an industry. This did not account for the subsistence hunts for whales, which was, for example in the Arctic, conducted by indigenous inhabitants, using indigenous technologies, and using the whale products directly within their communities without a chain of commerce involved in it. This type of whaling was of a fundamentally different character than the commercial whale hunt and was furthermore still actively ongoing during that time. After all, the colonisation of the Arctic and associated assimilation of the local indigenous population into the societal and economic constructs of the colonising states occurred slower than in other regions of the world. However, as I argue throughout this book, the consideration of indigenous whaling still being significantly different to other whaling is still an important factor in contemporary marine mammal law.

It would appear reasonable to assume that the 1931 whaling convention was applicable to states only that were actively engaged in whaling activities. This, however, was not the case and the convention was open for all states. This led to the joining of states with no immediate whaling interest, such as Poland, Serbia or Switzerland. This, in turn, led to an imbalance in the focal points of the convention's parties, particularly as regards conservation efforts: while there were many parties to the convention, only a few had a genuine interest in conserving whale stocks and it was left to them to implement conservation efforts. Moreover, many parties considered the provisions of the convention insufficient and already during the first years of its existence it became clear that this treaty was not adequate to conserve whales and thus ensure the future of the whaling industry, and to satisfy all parties concerned.

5.2.2 The 1937 Agreement for the Regulation of Whaling and Its 1938 Protocol

In light of the dissatisfaction that accompanied the 1931 whaling convention it became clear that new, different regulations were needed. This was particularly the case since both Germany and Japan started to actively whale in the Antarctic and bilateral agreements, particularly between Norwegian and British companies, were not kept. Consequently, towards the whaling season 1936/37 which was characterised by a chaotic landscape of bilateral agreements between companies and governments, it was clear that a new form of international regulatory regime was needed to regulate whaling activities worldwide, but particularly in Antarctic waters. Even though initially being sceptical of an international regulatory regime, it was the British government which convened a conference in London in the spring of 1937 in order to find a solution for a functioning regime on whaling. The number of states attending the conference was limited, however, reflecting upon the British-Norwegian dominance in the whaling issue, since it was Norway as well as the British Dominions New Zealand, the Republic of Ireland (Eire), Australia, and the Union of South Africa that attended. Moreover, Argentina and the United States, the latter increasingly pushing for a higher level of conservation in a new whaling regime, as well as Germany and Japan were invited.

It is important to remember that in 1937 the world stood at the brink of war. Therefore, the British government was hesitant to invite the US due to its conservation-based agenda, as well as Germany and Japan, but in the end issued formal invitations. While the US and Germany accepted the invitation, Japan did not. In fact, Japan openly dismissed the invitation and expanded its pelagic whaling programme in the same year. Also earlier attempts to either encourage or to coerce Japan to join the 1931 convention had failed. Japan's opposition to attend the 1937 conference also stemmed from the fact that it criticised British dominance and was in the process of expanding its whaling industry anyway. And, after all, one of the purposes of the conference was to find a way to limit Japan's expanding whaling operations in the Antarctic.[2] Japan therefore acted in pursuit of its own larger geopolitical and economic interests.

The nations that did attend the conference were once again divided into different groups whose interests needed to be reconciled. On the one side there were those nations that primarily focused on processing whales at sea in the rather newly-developed factory ships. These conducted whaling primarily in the Antarctic and

[2] Dorsey (2013), p. 66.

were the UK, Norway, and Germany. Also Japan belonged to this group of states which processed whales first and foremost for their oil. On the other side were whaling nations who focused on processing in land stations, such as Australia, New Zealand, Canada (for Newfoundland), and South Africa, primarily for whale meat. In addition, even though Norway and Japan put a strong focus on pelagic whaling in the Antarctic, they nevertheless also maintained land stations for meat, while also states such as Madagascar or the Faroe Islands engaged in shore-based whaling. In other words, the interests of whaling nations were not in unison, but the oil/meat—pelagic/terrestrial dichotomy needed to find adequate and equal reflection in a new whaling regime.

On 8 June 1937 the conference was adjourned and the attending nations had adopted the International Agreement for the Regulation of Whaling.[3] This agreement was to be in force only for 1 year unless the majority of its parties, which had to include Norway, the UK and Germany, favoured an extension of its duration. The Agreement mirrored some provisions of the 1931 whaling convention but it also went significantly further. First of all, the agreement applied to both pelagic whaling and land stations equally. Apart from right whales, also gray whales were to be fully protected while blue, fin, humpback and sperm whales had to reach a minimum size before they could be killed. Interestingly, the agreement now also included the taking of whales for scientific purposes. Scientific whaling was not considered under the 1931 convention, but found recognition in the US version of the convention upon its ratification. Now, scientific whaling had entered the international discourse on whaling. At that time, the drafters of the agreement could not know that scientific whaling would constitute a source of major disagreements in the future, as we will see below. But be that as it may, the agreement furthermore established whaling seasons and geographical boundaries for Antarctic whaling operations while also land stations and associated whale catchers were no longer allowed to whale longer than 6 months in the same area. Additionally, the wastefulness of whaling had emerged on the international agenda since based on the agreement, the whale carcass was to be fully used while whalers were required to take only as many whales as both factory ships and land stations were capable of processing.

In light of the short duration of the agreement, the parties decided to convene a second conference in order to prolong its lifetime. In June 1938, therefore, they came together in both Oslo and London, also inviting Japan, France and Denmark while Portugal attended as an observer. In order for the other states to join the agreement, it was amended through a Protocol, particularly aiming at incorporating the interests of the three non-parties. While France and Denmark signed the Protocol, Japan did not since it had too many reservations towards the 1937 convention. However, at yet another meeting in August 1939, Japan announced it would sign the convention for the next season. But with the outbreak of the Second World War, these plans did not materialise and the efforts to successfully implement the convention fell off the table.

---

[3]International Agreement for the Regulation of Whaling of 8 June 1937 (190 LNTS 79).

5.2.3 The 1946 Convention for the Regulation of Whaling

The regulation of whaling was understandably not a primary concern of the international community during the Second World War. In fact, the UK considered withdrawing from the 1931 convention and the 1937 agreement during that time in order to maximise oil production to fuel the war machinery. Only towards the end of 1942, when it became clear that German defeat would be just a matter of time, deliberations in the UK on a postwar regime for the regulation of whaling started. Although British and Norwegian whaling vessels were sunk during the war, the numbers were not so grave as to cripple the whaling industry. This was the opposite in Japan where it is estimated that around 95% of the whaling tonnage had been sunk while Antarctic whaling grounds could no longer be reached due to the dangers of naval warfare. The food crisis in Japan, for which whales were used as remedies, was therefore severe, which prompted the occupation forces in postwar Japan to particularly stimulate the whaling industry again. General MacArthur made Japanese whaling and Japanese international cooperation in this regard a primary issue. As a result, already in 1946 two remodelled oil tankers that served as factory ships were sent to the Antarctic in their pursuit of whales.

This notwithstanding, the first conference for a new regime on whaling was held in January 1944 in London, excluding Japan. Particularly the UK, Norway and the United States were the main players, the latter striving for increased conservation efforts for whales, the two former aiming to revitalise the slowed-down whaling industry directly after the war. The result was a short-lived and somewhat obscure protocol to which no primary information is available. However, since the United States had both the conservation of whales and the protection of the whaling industry in mind, at this 1944 conference, the US first proposed the establishment of a permanent whaling commission that would provide recommendations on catch quotas, whaling seasons, minimum lengths of whales or sanctuaries. In order for such commission to come to life, however, short-lived protocols to the 1937 convention were not feasible and the need for a completely new regime emerged. Before that, several countries, including the UK, Norway and the US, along with Mexico, France and some Commonwealth governments, agreed on yet another Protocol Amending The 1937 International Agreement For The Regulation Of Whaling at a conference in November 1945, again in London. Once again, the protocol was only applicable for one season, 1946/47, and did not establish a permanent commission nor did it find means and ways for permanent measures for the regulation of whaling. Even though the Bureau on Whaling Statistics was to be provided with data, this did not provide the Bureau itself with any regulatory competence.

Although the November conference resulted in a protocol, the United States were not satisfied with such a temporary arrangement and further pushed for the idea of a permanent commission. While originally the US idea was to create a regime which allowed merely trade between its members, the invitations to a conference to be held in December 1946, which were sent out to 18 nations, did not provide for any indication that trade with non-members would be prohibited. The planned establishment of the International Trade Organization, as part of the Bretton Woods institutions, thus placed the US in the position of having to gauge the conservation of whales and free trade in the postwar world. Disagreement, particularly from the Norwegian and British side, over the planned conference ensued about the invitation to states that had no previous history in whaling or that had not been parties to the 1930s agreements. The Netherlands, for instance, were such a case, but as the Dutch delegate at the conference noted, the Netherlands were forced to enter the whaling business after the war due to the shortage of oils and fats.

The United States proposed the establishment of an International Whaling Commission which would be following the *Zeitgeist* of 1945 as well as include somewhat progressive elements. The International Whaling Commission's overall task would be to provide guidance on the way whaling was to be conducted and which types of whales could be taken. Based on scientific advice, this guidance would not be static, but adjusted to changing conditions. This adjustment, however, could only be taken by a three-quarter majority of Commission members. These, however, would have the right to object to any decision and would thus not be bound to it. Even though the objection-clause did not find much appreciation among British and Norwegian delegates since it would potentially undermine the regulatory competence of the Commission, other states expressed their support for this approach since this would underline a state's sovereign right *not* to follow the decisions by the Commission. Sovereignty was also a primary reason for the dismissal of the US proposal to place the whaling commission under the Food and Agriculture Organization (FAO) of the United Nations. The UK and Norway considered their primacy in the whaling industry potentially undermined while the Soviet Union remarked that it is not part of the FAO in the first place.

The conference, which was held in November and December 1946, was marked by heated debates over the issues above, but resulted in the adoption of two distinct instruments: on the one hand, a new protocol for the 1937 Agreement was negotiated that would take effect for the 1947/48 season. On the other, the International Convention for the Regulation of Whaling (ICRW) was adopted, which was to take effect from the 1948/49 season onwards. With the adoption of the ICRW, also the International Whaling Commission (IWC) was established with its headquarters in London (relocated to Cambridge in the 1970s). The ICRW entered into force on 10 November 1948 and was ratified upon its coming into force by Australia, France, Iceland, Netherlands, Norway, Panama, South Africa, the Soviet Union, the UK and the United States. Japan ratified the ICRW and thus joined the IWC in 1951. On 17 December 2018 Japan announced its withdrawal.

With the ICRW in force, in August 1952 Chile, Peru and Ecuador signed a regional agreement for the South Pacific, the Regulations for Maritime Hunting Operations in the Waters of the South Pacific.[4] While the title of the agreement does not stipulate the fact that it refers to whales, the body of these regulation set specific rules for the South Pacific, implementing the ICRW in a South Pacific context. At the

[4]Regulations for Maritime Hunting Operations in the Waters of the South Pacific, 18 Aug 1952.

time of conclusion of these regulations, all three countries had been actively engaged in the pursuit of whales. Over time, and in particular in the 1970s to 1980s, the countries ultimately changed their position on whaling and up to this day have developed an outspoken anti-whaling position. Together with other Latin American members of the IWC, Peru, Chile and Ecuador are now part of the so-called 'Buenos Aires Group' which has streamlined its positions on whaling and which puts forth strictly conservationist proposals at Commission meetings. It is hard to pinpoint when the Buenos Aires Group was officially launched but official documentation of the IWC reveals the first occurrence of the term in 2008. It is therefore reasonable to assume that the group had formed some time in the early twenty-first century. While the Regulations are legally speaking still in force, given that the three countries are no longer engaged in whaling, they have become obsolete.

5.3 The International Whaling Commission, Whaling and Whales

The International Whaling Commission

- Based on the 1946 International Convention for the Regulation of Whaling
- Currently 88 member states
- Adopted a ban on commercial whaling in 1982 (in effect since 1985/86)
- Allows for whaling for scientific purposes
- Allows for non-commercial whaling conducted by indigenous populations
- Has been marked by a long-lasting dispute between pro- and anti-whaling countries

The International Whaling Commission came to life with the adoption of the International Convention for the Regulation of Whaling. In its early stages the IWC was labelled a 'whalers' club' due to the organisation's explicit focus safeguarding the whaling industry and not whales for their own sake.

5.3.1 Which Whales Are Regulated by the IWC?

In order to understand the IWC a bit better, we now turn to some fundamental elements that have shaped the past and are still shaping the present of the Commission. Let us therefore start with the most obvious element of the Commission: whales. As we have learnt in Chap. 2, there are around 80–90 different species of cetaceans in the world which are not all managed by the IWC. Instead, the Commission regulates the hunt for the 'great whales', which are more precisely being laid out in the Schedule of the Convention. In essence, however, this means that the Commission regulates whaling of all baleen whales as well as one toothed whale– the sperm whale. Dolphins species, including the orca, and porpoises, do thus not fall under the purview of the Commission. And it is also noteworthy that 'great' whales is more of a normative than size-dependent term: after all, some toothed whales, such as the bottlenose whale (*Hyperoodon ampullatus*) or the Baird's beaked whale (*Berardius bairdii*) either reach or even exceed the size of different minke whale species while a rather large number of toothed whales also exceed the size of pygmy right whales (*Caperea marginata*). Irrespective of size, neither the Convention nor its Schedule define the term 'whale'. Instead, the Schedule defines 'baleen' and 'toothed' whale, and explicitly sets catch limits for baleen whales and all the species it defines as being included in the term. It also explicitly sets catch limits for sperm whales. Hence, it does not set catch limits for toothed whales or other small cetaceans.

Bearing the above in mind, it is consequently incorrect to assume that the IWC is the watchdog over all whaling activities worldwide and that all whales are protected from hunting based on this organisation. Small cetaceans, killer whales and other dolphin and whale species fall under other regimes that regulate either their take or provide means for their protection. First and foremost, however, it is national legislation which regulates the taking of small cetaceans. A good example in this regard is the case of the critically endangered vaquita porpoise (*Phocoena sinus*) whose numbers range below 30 in the Baja California and whose numbers have dramatically declined due to the effects of by-catch in the Mexican totoaba (*Totoaba macdonaldi*) fisheries. Even though the issue of the vaquita has surfaced on the IWC agenda and a working group has been established to provide information and support for the species' protection, it lies primarily with Mexico to prevent the vaquita from going extinct.

5.3.2 The Organisational Set-up of the IWC

The IWC is an organisation that currently has 88 member states. This means that there are 88 states which are bound by the decisions taken by the IWC. All other states, in principle, can go whaling unimpeded and based on their own data and regulations. The membership of the IWC has not increased steadily, however, but has undergone fluctuations. Only towards the 1980s, a sharp rise in membership can be noted, followed by an even sharper rise in the new century (see Fig. 5.1).

While at first ten states ratified the convention and thus were IWC members, up until the late 1970s this number did not exceed twenty and also fluctuated quite notably. This means that states joined the IWC and then withdrew and possibly rejoined—such as countries like the Netherlands, New Zealand, Brazil or Iceland. At the same time, other states joined, so that the trend of membership constantly went up. A turning point was the time in the early 1980s when membership almost doubled within just a few years. This was the time when the so-called whaling moratorium was adopted, which we will get to know better below. Similarly, in the early 2000s towards 2019, membership increased dramatically with Liberia being the latest country to join the IWC in 2018. Thus, once the IWC establishes certain binding rules upon its members, these are contractually bound to them. Countries

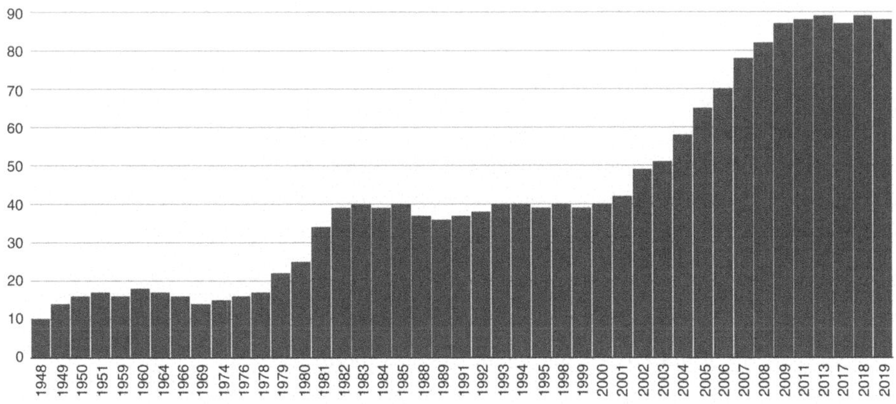

Fig. 5.1 Membership of the International Whaling Commission © Nikolas Sellheim

whose citizens go after whales that would fall under the purview of the Commission, but who are not member states are thus not bound to IWC rules. A good example in this regard is Indonesia, where small numbers of sperm whales are taken each year on the small island of Lembata. Also blue whale migration takes place in the waters surrounding the island, but the islanders consider them sacred and thus do not engage in any hunt. The point to make here is that even though the IWC has had strict no-take quotas for blue whales in place, it is up to the Indonesian government, and not the IWC, to allow or prohibit the Lembatan hunts since Indonesia is not a member of the Commission.

The IWC itself is an organisation whose mandate is to "(a) encourage, recommend, or if necessary, organize studies and investigations relating to whales and whaling; (b) collect and analyze statistical information concerning the current condition and trend of the whale stocks and the effects of whaling activities thereon; (c) study, appraise, and disseminate information concerning methods of maintaining and increasing the populations of whale stocks."[5] In order to do so, since 2012 the IWC meets every 2 years to discuss the way forward. Prior to 2012, meetings were held on an annual basis, but in order to reduce costs particularly for developing states, meetings are now conducted biannually.

The work of the Commission is conducted by several committees and working groups. Permanent committees are the Scientific Committee, which provides for scientific advice; the Finance and Administration Committee; the controversial Conservation Committee (see Sect. 5.3.4); the Aboriginal Subsistence Whaling Sub-Committee; the Infractions Sub-Committee; and the Working Group on Whale Killing Methods and Welfare Issues. Apart from the Commission meetings themselves, these committees have developed their own rules of procedure and have generated different subgroups and working groups.

---

[5] ICRW, art. IV.

5.3.3 How Is Whaling Regulated by the IWC?

The regulatory framework for the hunt for whales rests in the Schedule to the convention. The Schedule is an additional legally-binding document integral to the convention, which provides more detailed rules for member states to implement the ICRW. Since the Commission's inception, the Schedule has undergone numerous changes that have provided more and more shifts towards diverse and science-based decision-making. In principle, the Schedule sets quotas for specific whale species in specific regions, theoretically allowing member states to hunt these whales for commercial purposes. The Scientific Committee, which has been part and parcel of the Commission's decision-making, provides the necessary scientific input that ensures both the longevity of the whale species themselves and, at the same time, their hunt. This idea of 'sustainable use' of whales is consequently a part of the *raison d'être* of the Commission from the very beginning.

While nowadays quotas determine the numbers of whales that can, or cannot, be caught, this system was introduced only in 1972 when the so-called blue whale unit (BWU) was abandoned. The BWU was a system that was initially used by whaling companies to regulate their whaling activities, but was taken over by the IWC upon its inception. Based on this idea, countries were allocated specific BWUs based on which they were able to hunt whales on a ratio-basis: one blue whale for two fin whales, or two and a half humpback whales, or six sei whales. This strictly efficiency-based approach led to the drastic overhunting of large whale species such as the blue whales. In other words, the more large whales could be caught, the faster the BWU was reached and the oil could be placed on the international market. This led to what has been coined the 'Whaling Olympics' in which large whale species were viciously overhunted for commercial gain, and despite the existence of the regulatory body IWC.

From 1972 onwards, the issue of declining whale populations had also entered the overall discourse on environmental protection, fuelling the environmental movement, which emerged during the 1960s. Indeed, also the United Nations recognised the need for increased international efforts to protect the natural world. In order to develop international means to safeguard the environment, the first UN Conference on Environment and Development was held in Stockholm in June 1972. This watershed event marked the birth of global initiatives that, in the longer run, resulted in the establishment of more nuanced institutions and regimes dealing with environmental protection. As an outcome of this conference, also the United Nations Environment Programme (UNEP) was established with its headquarters in Nairobi, Kenya. Furthermore, it was also at the Stockholm conference when particularly the United States pushed for a global moratorium on commercial whaling in order for the ever-decreasing stocks of large whales to recover. Since the conference itself did not take any legally-binding decisions, but rather provided for a roadmap for the future, it could not force the attending states—or the IWC, for that matter—to impose a moratorium on whaling. Instead, the conference adopted an Action Plan for the Human Environment in which the IWC is called upon to establish a 10-year moratorium on commercial whaling. What that meant was that normative pressure was exerted on the IWC to have its member states alter the Schedule in a way that allowed for a temporary, 10-year zero-catch quota for all whaling that is being conducted for commercial purposes. While not directly mentioning it, this proposal did not account for aboriginal whalers whose subsistence whaling activities did and do not fall under the category of 'commercial' whaling.

In order for a Schedule amendment to occur, the Commission's member states need to provide for a three-quarter majority. Any proposed amendment that does not meet this criteria is not adopted and thus does not become binding upon member states. In the post-Stockholm 1970s, this majority could not be reached and it was particularly whaling states such as Japan, Norway and Iceland that strongly opposed such a moratorium. Since the IWC is an organisation which is open to all states, the shift towards a pro-moratorium three-quarter majority was reached upon the joining of an increasing number of states opposing whaling on normative grounds, which occurred towards the early 1980s. By 1982, a sufficient numbers of states within the IWC supported a moratorium and the IWC amended the catch quotas in the Schedule to be zero for 10 years. In this sense, the IWC did not explicitly outlaw whaling as such, but rather established a (temporary) zero-catch-quota for all whale species listed in the Schedule.

The IWC does not 'police' its own rules, however. Consequently, if a state does not conform to IWC rules, there is no way of enforcing the Commission's decisions or its Schedule. The only way to monitor whether or not rules are being followed is through self-reporting to the Infractions Committee which records infractions and suggests how to proceed in a particular case. However, it is then the Commission itself which decides upon certain sanctions on member states that do not comply with IWC rules. What the nature of these sanctions in relation to illegal hunting of whales are is difficult to assess. It is mostly the government of the infracting state which is to take punitive action against those having committed the infraction. In case of the government sanctioning the infraction, the IWC becomes powerless, which has ascribed many to label it as an organisation with 'no teeth'. Indeed, in case of the Soviet Union, in the 1960s and 1970s, Soviet whalers reported wrong data to the Scientific Committee, which was unable to independently verify this data. The IWC in turn was unable to sanction the Soviet Union, making the organisation an incapable bystander in watching whale populations drop.

The most commonly found non-compliance with IWC rules occurs with regard to financial contributions. If a member state does not pay its dues to the Commission, it can be sanctioned. These sanctions have direct effect and may occur in the form of suspensions of voting rights or the non-provision of specific services from the IWC. The IWC is not able to expel a member due to unpaid fees or non-compliance with other provisions.

5.3.4 The Moratorium on Commercial Whaling: A Turning Point at the IWC

The IWC is often called an organisation that is marked by dysfunctionality or as being in a deadlock. Indeed, it is deeply divided between states favouring the sustainable use of whales and those states favouring the conservation of whales. But as we have seen above, whaling has always been a divisive activity. Either on the question of whether an international regulatory body is necessary in the first place; or on how to integrate the different types of whaling under one umbrella. While all other divides were eventually overcome, the question of sustainable use vs conservation appears to be irreconcilable. The major turning point for the IWC was the passing of the moratorium, or the zero-catch quota, at the 1982 meeting in Brighton, England, when 25 states voted in favour, 8 against with 5 abstaining for the respective Schedule amendment.

Throughout the 1970s, the divide on how to approach whaling had grown more solid. On the one hand, whaling states did not wish to put a blanket moratorium in place if scientific findings did allow for some form of commercial whaling. Or put differently, a moratorium was considered not justifiable from a scientific standpoint and that, as the Scientific Committee expressed in 1973, a "blanket moratorium is in the same category as a blue whale unit quota, in that they are both attempts to regulate several stocks as one group whereas prudent management requires regulation of the stocks individually."[6] On the other hand, the political and public discourse established 'the whale' as a type of species that is so special that any form of utilisation is no longer acceptable. Essentially, throughout the 1970s the IWC had to decide which way to go forward since more and more states joined the IWC that were either actively whaling or that had a whaling history. The joining of Sweden in 1979 marked a breaking point since the country was the first one to join the IWC with the purpose of voicing its opposition to commercial whaling for principal reasons. From then on, the number of countries joining the IWC increasingly opposed commercial whaling, which in the end led to the three-quarter majority of states supporting the putting into place of a zero-catch-quota.

The questions that the Commission dealt with were essentially of a threefold character: (1) is the killing of whales ethically justifiable given that they might be sentient beings? (2) If it is, is there sufficient scientific evidence to ensure that whales are being hunted sustainably? (3) Does the Commission have the right to prevent states from using their own resources? In 1980, particularly Australia stood at the fore in the question of a global ban on commercial whaling, both from a scientific perspective as well as due to principal considerations. Also the United States and the Seychelles were champions of a moratorium on whaling, however, contrary to Australia, the latter countries considered this a temporary moratorium while Australia's approach did not hold temporary elements.

[6]IWC (1973), p. 38.

At the core of the scientific discussions surrounding a moratorium stood the so-called New Management Procedure (NMP). This procedure was developed by the Scientific Committee and aimed at increasing whale stocks until they reach their maximum sustainable yield (MSY), determining the maximum numbers of whales that can be killed without compromising a whale stock. But from the very beginning, states questioned the reliability of the NMP and preferred not to kill whales at all in order for them to recover instead of running the risk of faulty science. Also other considerations, that went hand in hand with the changing discourse on whales, naturally fed into the deliberations at the IWC. We must remember that towards the end of the 1970s the environmental and animal rights movement had gained significant support and had grown steadily—to a degree that allowed them to be present at IWC meetings and to conduct effective lobbying. 'The whale' served an important function as a poster species, the protection of which was a core demand from supporters of anti-whaling campaigns in the West. Moreover, environmental and humane organisations developed expertise that they were able to demonstrate at IWC meetings, either by providing delegations with scientific data or with proposals that delegates would then present in the plenary for discussion. Over the course of time, representatives of environmental organisations would furthermore become parts of delegations—a practice that persists to this day. The same is also true for delegations in favour of whaling where industry representatives are part of the country delegations of Japan or Iceland, for instance. Again, this practice is not from the past, but is still prominent in ongoing meetings. Moreover, it is a common sight for the observer of the IWC to see country delegates and NGO/industry representatives either dining together or holding closed meetings.

But be that as it may, upon the early years of the 1980s, a moratorium on all commercial whaling moved closer to become a reality. The first step was taken in 1979 when a zero-catch quota was agreed upon for factory ships, except for those hunting minke whales. The motivation behind this move was that all in all factory-ship whaling was considered more detrimental than land station whaling. But since it was not clear when this factory-ship moratorium would come into effect, particularly Soviet whaling fleets continued hunting several hundred sperm whales during the 1979/80 season in the North Pacific. In combination with doubts over the NMP, the calls for a complete moratorium on commercial whaling grew stronger and more concrete in 1981. By that time, Canada was the first country to leave the IWC due to the organisation's stance on commercial whaling. Canada feared that the IWC would infringe upon state sovereignty on the high seas, which was to be manifested in the 1982 UN Convention on the Law of the Sea (UNCLOS) and the newly established 200-mile Exclusive Economic Zone (EEZ). Moreover, Canada was concerned that the IWC would take control over its Arctic narwhal, beluga and right whale hunts, conducted by Inuit and other indigenous communities. The United States, the United Kingdom, the Seychelles and France, along with Australia, New Zealand, Sweden and others, on the other hand, were vocally in favour of a blanket moratorium. At the core of their arguments stood three main concerns: the past failure of the IWC to effectively protect whale stocks; uncertainties about stock assessment methods and the determination of MSYs; and concerns over the appropriateness of whale killing methods. While these were all essentially scientific concerns, also a principal opposition to whaling cannot be neglected, as mentioned above.

With the deepening discussions on whether or not a moratorium would be put in place, also the rift between members of the IWC became more obvious. While issues related to science, sovereignty and ethics are products of this discussion, the proposal also has a normative meaning for the Commission. Of course, disagreements over the whaling issue were and still are the driving point, but also in other organisations disagreements exist which do not automatically lead to a deadlock as in the IWC. This deadlock is more rooted in the overall role and function of the Commission. Or in other words: is the role of the Commission to protect *whales* or to protect and regulate the *whaling industry*? Of course, those favouring whaling note that it is the latter while those that want to halt whaling support the former view. This issue boils down to the text of the Convention itself, its objective and how different states interpret this objective. Particularly Japan has argued that a moratorium is in contravention of the Commission's objective, which it sees as "conservation, development, and optimum utilization of the whale resource", as per article V (2) of the ICRW. This means, in Japan's view, that both the conservation *and* optimum utilisation (sustainable use) are the key roles of the Commission. At the centre of this should stand scientific knowledge and advice. The Scientific Committee concurred with this view. It is therefore not surprising that upon the adoption of the moratorium, Japan filed an objection, meaning that the moratorium did not become binding upon it. The same accounted for Norway, Peru and the USSR. However, while Japan and Peru had withdrawn their objections by 1987, Norway and the USSR (now Russian Federation) have up to this day not done so. Therefore, Norway and Russia can still conduct commercial whaling in spite of the zero catch quota. The withdrawal of Japan's objection appears surprising. However, this occurred due to economic pressure from the United States.

Scientific Uncertainty and the Precautionary Approach

The RMP is fundamentally rooted in the integration of scientific uncertainty or lack of data in the determination of catch limits. The algorithm that has developed out of this work is referred to as the Catch Limit Algorithm (CLA). In order to apply the CLA, two key pieces of information are to be provided: first, abundance estimates at 6-year intervals, including statistical uncertainty. Second, estimates of past catches as well as current catch levels. With this information, the CLA can be determined.

As a general rule, the CLA as applied by the IWC aims to maintain a minimum population of 54% of the pre-catch population while catch limits are set in a way that the population stabilises at 72% of pre-exploitation levels. The Commission opted for 72% levels as one of the options presented by the Scientific Committee, providing for high conservation objectives and lower catches. However, the Scientific Committee also presented other levels,

(continued)

i.e. 60% with higher catches and lower conservation; and 65% as an intermediate level. These three options are referred to as 'tuning levels': 0.60, 0.65 and 0.72. All of these levels, despite their differences in conservation and catch outcomes, are guided by the precautionary principle.

At the 67th meeting of the IWC in 2018, tuning levels also played a role. I observed how Iceland was accused of applying 'conservative' tuning levels for its whaling quota which were deemed to be unsustainable. Iceland, however, countered this accusation by referring to the tuning levels of the Scientific Committee which it adheres to. We see therefore that although the Scientific Committee did indeed provide for these tuning levels, their application is a matter of political (and arguably emotional) decisions.

Others, however, place the focus of the IWC on the conservation of whales and not so much on their utilisation and the development of the whaling industry or whale resource. At the core of this argument stands the contextual consideration of the whaling convention. It is argued that since the whaling industry is no longer of such influence as at the time of the ICRW's conclusion, the political, social and ultimately legal contexts have changed. Paired with the overall public discourse on environmental protection, the IWC has thus shifted to become a more conservation-oriented organisation than one dealing with the management of whaling. And here lies the crux of the issue: since there is, in this regard, not a right or wrong way to interpret the Convention, there is no way to resolve this dispute. Several attempts have been made to 'normalise' or 'modernise' the IWC—to find means and ways to overcome this rift. But up to this day, these attempts have failed.

Instead, the IWC has sharpened its conservation-oriented agenda, much to the dismay of Japan and other whaling countries. Since it is necessary to change the Schedule based on a three-three-quarter majority, Japan has continuously attempted to get states to join the Commission and support its cause. To this end, there is evidence that Japan has 'weaponised' its overseas development aid for countries in the Caribbean and South Pacific, meaning that funds to support local projects were paid in return for a country joining the IWC and voting in favour of Japan's pro-whaling stance. The degree to which this has happened is difficult to assess, but that fact that it has indeed happened is indisputable.[7] At the same time, more and more countries have joined the IWC that either have no whaling history and/or that are explicitly anti-whaling. In how far rich anti-whaling countries have used political or economic means to lure other states to join the Commission has, at least to my knowledge, not been researched so far. Since the convention itself does not contain a mechanism to deal with such a development—after all, the drafters of the convention could not foresee what would happen—the Commission has reached a stage which

[7] Miller and Dolšak (2007).

makes effective decision-making in regard to whaling difficult and certainly not consensus-driven.

This is especially the case since the moratorium on commercial whaling was to be temporary from the outset. First of all, between the passing of the moratorium and its coming into effect lay almost 4 years since it was the 1985/86 whaling season that saw the moratorium applied. Secondly, the moratorium was to last for 4 years until the Commission was to conduct a review of its effectiveness in 1990. This review, however, did not materialise in the way it was supposed to: implying a review with the intention of potentially lifting the moratorium. Instead, the majority of Commission members reaffirmed the moratorium, particularly in light of the decreasing importance of whaling and scientific uncertainty. Indeed, upon passing of the moratorium, the Scientific Committee was tasked to develop a new method of determining whale stocks and their exploitation, bearing in mind the shortcomings and concerns related to the New Management Procedure of the 1970s. This Revised Management Procedure (RMP) was adopted by the Commission in 1994 and provided a new methodology for the assessment and possible utilisation of whale stocks. The specificity of the RMP lay in its inclusion of scientific uncertainty and in the role of cetaceans in the marine ecosystem.

With the adoption of the RMP, the IWC accepted a scientific means to determine the exploitability of whale stocks and also accepted the scientific basis for a lifting of the moratorium. However, before a lifting of the moratorium was to be discussed, the Commission was to furthermore agree on a scheme that allowed for observation and for a system that insured that quotas were kept. The RMP was therefore to be complemented by non-scientific elements, both of which were to accumulate in the Revised Management Scheme (RMS), which the Commission agreed to work on and ultimately adopt. However, since the adoption of the RMP, Commission members have not managed to agree on the managerial aspects of the RMS, particularly as regards the inclusion of a binding code of conduct for scientific whaling. While some countries pressed for such code, others considered it premature, while again others urged for the amendment of the Convention in order to do away with scientific whaling altogether. In 2006, the Commission agreed that it had reached an impasse on the discussions surrounding the RMS. This resulted in the *de facto* staying in place of the moratorium up to this day.

The RMS plays an important role in Iceland's position within the IWC. Iceland left the IWC in 1992 and did so due to the moratorium on commercial whaling. Even though Iceland was the first country to conduct scientific whaling in its own waters, which was in accordance with the convention, as we will see in the following section, it sold the meat to Japan, thus effectively circumventing the ban on commercial whaling. Not surprisingly, this caused significant dissatisfaction amongst Commission members and civil society organisations. Although Iceland voted against the moratorium, it did not formally object to it and was therefore bound to it. Along with Norway, Greenland and the Faroe Islands, in 1992 Iceland established the North Atlantic Marine Mammal Commission (NAMMCO), which promotes the conservation and sustainable use of marine mammals (see Sect. 4.4.2). Iceland therefore did not consider staying in the IWC a viable option anymore. In the course of the 1990s, however, Iceland reconsidered its position in the spirit of cooperation relating to the global commons and rejoined the IWC in 2002. Much to the dismay of many anti-whaling countries, Iceland voiced its reservation to the moratorium and linked its whaling activities to the adoption of the RMS: it would not authorise whaling while progress is being made on the negotiations of the RMS. Essentially, therefore, Iceland rejoined the Commission under the assumption (or expectation) that some form of commercial whaling would be allowed under the RMS. However, since no progress is being made, Iceland resumed commercial whaling in 2006.

Over the years, Japan and other whaling nations have continuously pushed for a lifting of the moratorium, but the IWC has continuously moved towards a more stringent conservation-oriented agenda. In 2003 this was manifested in the establishment of the Conservation Committee, whose mandate falls under the strengthening of the conservation agenda, as was stipulated in the respective Resolution 2003-I *The Berlin Initiative on Strengthening the Conservation Agenda of the International Whaling Commission*. Not surprisingly, Japan, Norway, Iceland and some other states have refused to endorse the resolution and associated establishment of the Conservation Committee and have refrained from supporting it financially. Instead, especially Japan has continuously tabled new proposals for resolutions or Schedule amendments, which would allow for a limited resumption of commercial whaling, thus ultimately lifting the moratorium. This has fallen on deaf ears for the majority of Commission members and Japan's proposals have failed. One of these long-term, unsuccessful attempts was the proposed introduction of a new category of whaling: small-type coastal whaling (STCW), which was to be placed somewhere between commercial whaling and Aboriginal Subsistence Whaling (ASW) in that it would provide for a small quota for Japanese coastal whalers. However, since also STCW would effectively mean a Schedule amendment which would undermine the zero-catch quota without the adoption of the RMS, these proposals have not been adopted by the Commission.

The last of Japan's attempts to lift the moratorium was tabled in 2018 in Florianópolis, Brazil. Here, Japan proposed a co-existence of different views within the Commission and, at the same time, an associated Schedule amendment which would provide Japan with a very limited minke whale quota for coastal whaling purposes. This stood in stark contrast to the Brazil-proposed Florianópolis Declaration, which saw the increased strengthening of the IWC's conservationist agenda. Japan's dual proposal failed while the Commission adopted the Florianópolis Declaration. Following the votes, heated debates ensued, which prompted Japan's State Minister of Agriculture, Forestry and Fisheries, Masaaki Taniai, to indicate Japan's withdrawal from the IWC by noting that in the coming months Japan had to conduct a "fundamental reassessment of its position as a member of the IWC."[8]

On 17 December 2018, Japan officially announced its withdrawal from the International Whaling Commission, coming into effect on 1 July 2019. This move enables Japan to conduct whaling in its EEZ. At the same time, Japan announced that

[8]Taniai (2018)

it would immediately abandon its scientific whaling activities in the Southern Ocean—themselves significant causes for international dissatisfaction.

5.3.5 Scientific Whaling: The Loophole?

Even though a zero catch quota has been in effect since 1985/86, thus banning any lethal use of whales listed in the Schedule for all those countries parties to the whaling convention, this does not account for whales killed for scientific purposes. This scientific whaling has legally been possible since the adoption of the ICRW in 1946 although it had not been actively pursued until the moratorium of 1982. As we have seen above, it was Iceland which in 1986 initiated the first scientific hunts for sei and fin whales in Icelandic waters. Since the sale of whale meat is not explicitly forbidden under this 'special permit' type of whaling, Iceland consequently did not act in contravention to the convention. In fact, although the Commission has put the moratorium in place, it is the nation states themselves that decide to issue permits for the killing, taking and treating of whales as they see fit for scientific purposes. Moreover, the convention stipulates that whales that are taken under a special permit are to be processed in a way the issuing government considers appropriate. This means that the government of Iceland issuing permits and then selling the produce stemming from these whales to Japan is, legally speaking, not illegal. What the convention is silent about is the question of how to deal with the sale of whale meat in light of a ban on commercial whaling. Hence, the conduct of scientific whaling—even though it may yield scientific data—may be conducted *in order to* sell the meat to other parties. This would ultimately jeopardise the purpose of putting a ban on commercial whaling in place in the first place and all countries could conduct scientific whaling, generate scientific data and then sell the meat based on its own preferences.

This problem has been labelled by many as the 'loophole' of scientific or special permit whaling under the ICRW. While strictly speaking perfectly legal, it constitutes a problem in light of the ban on commercial whaling. However, this loophole is only a loophole in this regard. If the RMS were in place and a limited commercial whale hunt were possible, the characterisation of the special permit clause being a loophole would consequently diminish since the demand for whale meat in a third country would be satisfied through official commercial pathways.

Iceland, however, is not the only country having conducted scientific whaling and an associated commercial component. In fact, Iceland has not conducted scientific whaling since its official resumption of commercial whaling in 2006. Instead, it has been Japan which has stood at the forefront of international controversy surrounding its special permit whaling. Up until its announcement of the abandoning of its Antarctic whaling operations, Japan had two branches of scientific whaling programmes in place: one in the North Pacific and one in the Southern Ocean, the latter of which has triggered the *Sea Shepherd Conservation Society* (SSCS) to actively confront Japanese whaling vessels and physically hinder them from conducting whaling. The TV show *Whale Wars* is a testament (and arguably a dramatisation) of the conflict in the Antarctic.

The administrative arm of Japan's research on whales is the Institute of Cetacean Research (ICR) in Tokyo, which came into being in 1987 and emerged out of the Whales Research Institute of 1947. Even though the ICR is supposedly a non-profit organisation, it receives funding from the Ministry of Agriculture, Fisheries and Forestry, whose Fisheries Agency represents Japan at the IWC. Moreover, there are claims that other funding for ICR's activities stems from profit-oriented players within the whaling industry.[9] These claims cannot be independently verified, however.

Be that as it may, Japan's whaling programmes have generated a wide array of scientific publications and the country has been an active contributor to the work of the Scientific Committee of the IWC. While as such Japan's research programmes are not in question, the necessity of killing whales to obtain research results is subject of the controversy. However, as the ICR makes unmistakably clear, the purpose of the research conducted particularly in the Antarctic is to show that the sustainable use of whales is feasible and possible. To this end, Japan considers it necessary to conduct lethal and non-lethal research.[10] In other words, if the research objective were a different one, lethal research may not be necessary. Or to put even more differently, if Japan were to be able to conduct commercial whaling elsewhere, there would not be a need to show that also in the Antarctic commercial whaling would be possible. Even though the official Japanese statement on Japan's withdrawal[11] merely mentions the cessation of Antarctic whaling, the fact that it now can commercially whale in its own territorial waters and EEZ makes Antarctic research whaling obsolete.

The question of whether or not scientific whaling is commercial whaling in disguise is not a mere academic question, but also one with legal significance. After all, the ICRW clearly allows whaling for "purposes of scientific research."[12] If whaling was conducted for other purposes, this would mean a breach of the convention's provision and would thus be illegal for those parties conducting such whaling. To fully understand this issue, however, we must now briefly turn to two other elements in the context of legality of Japanese Antarctic scientific whaling. The first one is that of whale sanctuaries. Whale sanctuaries are designated areas in which all forms of commercial whaling are prohibited, irrespective of whether a global moratorium is in place or not. The first sanctuary was established already in 1979, thus 3 years before the adoption of the moratorium, and encompasses the entire Indian Ocean up until 55° South latitude. The second was established in 1994 and is the one relevant for the controversies surrounding Japanese whaling in the Antarctic since it is called the Southern Ocean Whale Sanctuary (SOWS). It encircles the

[9]Morikawa (2013)

[10]ICR (2011).

[11]Prime Minister of Japan and his Cabinet (2018).

[12]ICRW, article VIII (1).

entire Antarctic continent, with its coordinates specifically outlined in the Schedule. Apart from these established sanctuaries, over a number of years, Brazil and other southern Atlantic littoral states have proposed the establishment of a South Atlantic Whale Sanctuary, but have thus far failed to garner the necessary three-quarter majority to amend the Schedule accordingly.

Not surprisingly, Japan has been an outspoken opponent of the establishment of whale sanctuaries since it considers these contrary to a science-based approach of the Commission. After all, whale sanctuaries establish blanket bans on the taking of whales in specific areas, neglecting differences in stocks and, ultimately, negating the migratory behaviour of many whale species. But be that as it may, despite the existence of the Southern Ocean Whale Sanctuary, Japan has continued whaling in it, since, officially, whaling operations were not for commercial but for scientific reasons.

Here we enter the second element of relevance in this context: Australia's Antarctic claims. Seven states claim territories in Antarctica—Australia, New Zealand, the UK, Chile, France, Argentina and Norway—some of which overlap, based on their historical activities on the continent. However, the Antarctic Treaty of 1959, which these states agreed on, but which has been joined by numerous other states, including Japan, since then, does not substantiate such claims as long as the treaty is in force. The treaty notes that "[n]o acts or activities taking place while the present Treaty is in force shall constitute a basis for asserting, supporting or denying a claim to territorial sovereignty in Antarctica or create any rights of sovereignty in Antarctica."[13] To this end, while all claimant states recognise each other's claims, other states do not. Japan, for example, does not recognise any Antarctic territorial claims.

Australia has time and again made reference to its claim, also by submitting information on its EEZ in its Antarctic territory to the UN Commission on the Limits of the Continental Shelf (CLCS) in 2004. According to the UNCLOS, the coastal state has sovereign rights over the marine resources within its EEZ, up to 200 nautical mile from the baseline. To this end, Australia retains the right to control any fishing or whaling activities in its EEZs, also in its Antarctic territory. Since Australia has long criticised Japan's scientific whaling operations in the Antarctic as being commercial whaling in disguise, it consequently considered Japan's whaling in its Antarctic EEZ as being in contravention of Australian law.

Australia took therefore an unusual step: it filed proceedings at the International Court of Justice (ICJ). The ICJ is an international institution which aims to resolve bilateral disputes. Even though whaling is indeed a multilateral activity, in the *Whaling in the Antarctic* case, it essentially boiled down to a bilateral dispute between Australia and Japan. Australia filed an application to start proceedings against Japan on 31 May 2010, based on the assertion that Japan's ongoing Antarctic whaling programme, JARPA-II, violated the moratorium on commercial whaling based on the ICRW and the associated Australian stance on whaling. The court, from

[13] Antarctic Treaty, art IV c).

the outset, was in a difficult position. After all, it would have to consider whether the Australian Antarctic Territory and associated EEZ would be an admissible legal basis even though many governments do not recognise Australian sovereignty. In order to avert any potential future dispute over this issue, the court decided not to deal with issues related to maritime delimitation—and thus not to accept Japan's argument to dismiss the proceedings due to the uncertain legal situation regarding sovereignty in Antarctica—but instead to focus on the larger picture of Japanese Antarctic whaling: namely, whether it is indeed in contravention of the moratorium and thus whether it is "for the purposes of scientific research", as the ICRW states. This would furthermore constitute a breach of the ban on whaling in the Southern Ocean Whale Sanctuary.

The court faced a dilemma, however: while its judges are experts in the field of legal sciences, they are not whale scientists and still had to determine whether Japan's scientific whaling programme corresponded to legitimate scientific standards or not. Or to put into simpler words: was Japanese Antarctic whaling 'science' or not? In order to circumvent this question, the court applied extremely complex and technical reasoning and focused on the question of whether whaling operations were 'for the purposes of' science and thus not on science itself. Thus, the court examined whether the lethal taking of whales in the Antarctic is appropriate for reaching the stated research objectives of JARPA-II. Fitzmaurice and Tamada have conducted a detailed analysis of the case, touching upon the multidimensional aspects of the reasoning of the different parties and the court.[14] Without needing to delve into the complexities of the case, in its judgement of 31 March 2014, the court found that Japan's whaling programme was not primarily conducted 'for the purposes of' scientific research and thus constituted a violation of the moratorium and of the ban on whaling in the SOWS. Japan was therefore ordered to revoke the special permits and to halt whaling in the Antarctic. While seemingly a victory for Australia and other anti-whaling states, the court did not order Japan to halt whaling in the Antarctic entirely. After all, the programme in question was merely JARPA-II, consequently enabling Japan to continue whaling with a different research programme and associated research objectives.

Upon issuing of the judgement, Japan immediately revoked its special permits and thus stopped JARPA-II. Yet, it also immediately started working on a new research programme, NEWREP-A. The programme saw the taking of 333 Antarctic minke whales annually for a period of 12 years for scientific purposes. Despite the proposal having been presented to the IWC's Scientific Committee, the Commission could not agree on whether or not it could endorse the programme. Disagreement existed over the necessity to take such a large number of whales for the objectives outlined in the programme. The Scientific Committee set out to review the programme at the request of the Commission, in order to determine whether the lethal takes were justifiable, but Japan did not wait and issued the permits before the

[14]Fitzmaurice and Tamada (2014).

review was complete. I cannot ascertain how many whales have finally been taken under NEWREP-A, but once Japan announced its withdrawal from the IWC in December 2018, it furthermore announced its cessation of scientific whaling in the Antarctic. Instead, it would now increasingly focus on scientific (and commercial) whaling in the North Pacific, the former to be conducted under the new research programme NEWREP-NP. In 2019 it was announced, however, that also North Pacific scientific whaling operations would be ceased by 2020. From then on, scientific research would be done based on the whales landed by the commercial whale hunts and predominantly based on non-lethal research practices.

5.3.6 Aboriginal Subsistence Whaling: The Question of Traditionality and Necessity

The scientific hunt of whales is, apart from commercial whaling, not the only controversial element within the IWC. The other major source of heated and emotional discussions at IWC meetings circles around Aboriginal Subsistence Whaling (ASW). From the early days of the Commission, it was particularly the United States that have pushed for an exemption of aboriginal whalers from any regulatory measures decided by the Commission. Already the Schedule amendment decided at the first meeting of the Commission in May 1949 included a provision that the killing of gray and right whales is forbidden unless they "are to be used exclusively for local consumption by the aborigines."[15] This exemption was further expanded the more regulatory steps the IWC took. In 1977, IWC reports make direct reference to aboriginal subsistence and the cultural needs of aboriginal peoples. In 1980, the term 'aboriginal subsistence whaling' occurs for the first time. During the discussion surrounding the moratorium it became quickly clear that aboriginal whaling and commercial whaling are two different types of whaling. While the latter aims for large numbers and ultimately for generating profit, the former is of a much smaller scale and serves primarily local needs. Or to turn it around, it is not primarily conducted for commercial gain but to directly feed the community it was hunted by. To this end, a moratorium was not to impact these hunts and the Schedule was amended accordingly. To further substantiate this difference, in 1983 the Commission adopted the Resolution on Aboriginal Subsistence Whaling, in which it pledged to recognise aboriginal needs and to include aboriginal peoples in the management processes.

Several elements coalesce in the consideration of ASW. First of all, a clear ethnic dimension is involved: whalers who fall under this category must be indigenous/native/aboriginal. Even though international law has not found a definition for these terms—and particularly not at the time of the conclusion of the ICRW or at the time of the moratorium—it is nevertheless rather clear that they have lived in the respective whaling area prior to the commencement of commercial whaling and

[15]ICRW, Schedule (1950), para. 2.

that they are ethnically different to the commercial whalers. Second, the hunt conducted by these peoples must strictly be non-commercial, i.e. for subsistence purposes. This means that the spheres of exchange of goods stemming from these hunts are not to generate profit and must not be market-oriented. It is thus the communities themselves which make use of the goods stemming from these hunts.

While theoretically this approach provides an avenue for aboriginal communities to have their needs recognised, in practice the story looks quite different. First of all, the idea of a purely community-based sphere of exchange is an idea which stems from the late nineteenth century when the first 'aboriginal exemptions' were introduced into seal management regimes (see Chap. 6). The idea was, of course, to differentiate between large-scale and small-scale hunts. Since small-scale hunters were colonised native communities, the ethnic dimension inevitably was an important marker. Particularly since other settlements in regions of whaling (and sealing), even though they might have conducted small-scale marine mammal hunts, were in all likelihood in one way or another attached to commercial hunting operations. Consequently, the focus was put on *aboriginal* subsistence hunting. However, as regards *subsistence* hunting, the perception was that these hunts were fundamentally different, and indeed separate, from the commercial hunts. Trade pathways, market considerations and other characteristics that dictated the numbers of animals that were taken were perceived as not playing a role in aboriginal subsistence hunts. Consequently, any regulatory regimes, including whaling regulations, should exempt these hunts from any restrictions.

While in the early days of colonisation, especially in the Arctic, this was certainly true, over time the two types of hunt began to intertwine increasingly. The more native communities became dependent on goods from their colonisers—both in good and bad ways—the more monetary considerations started to play a role. Hence, the idea of a 'noble savage' that does not engage in commercial hunting activities eroded. John Bockstoce treatise on the fur trade in the Western Arctic is a testament of this development.[16]

This meant in practice that aboriginal whalers increasingly resorted to the use of 'non-traditional' technology such as motorised boats or rifles, the purchase of which inevitably required money. Therefore, since the main sources for these goods were commercial marine mammal hunters, aboriginal whalers inevitably became attached to the commercial marine mammal hunt, while, additionally, also dietary habits and practices changed: canned and other commercial food products were introduced, effectively replacing—or at least diminishing—the consumption of 'traditional' foods such as whale or seal meat. The question thus arose whether aboriginal whalers would still have a need for whaling.

While until the adoption of the moratorium aboriginal needs did not play a major role in IWC deliberations, this changed dramatically when all commercial whaling was prohibited. While rather consensually the Commission agreed that aboriginal whalers should not be affected by the moratorium due to the small numbers of

[16]Bockstoce (2018).

whales they hunted, given the population status of right or gray whales, also aboriginal whalers could not continue as before. In order to continue whaling, the Commission had to amend the Schedule in a way that would afford aboriginal whalers a quota for specific species over a certain period of time. Before this could be done, however, it was up to the governments of those countries in which aboriginal subsistence whaling occurred to submit a 'Needs Statement' to the IWC in which it is clearly outlined why the respective indigenous people is in need of whale hunts. This practice continues to this day and up until IWC67 in 2018, every 6 years a new application for a new quota had to be filed. We will turn back to this below.

Currently, there are four countries in the IWC in which ASW is conducted: the United States for the Eskimo and Makah whale hunts; the Russian Federation for whaling in Chukotka; Denmark for whaling in Greenland; and St Vincent & the Grenadines for the Bequia humpback whale hunts. While the IWC had rather little concerns granting a quota for the Eskimo, Chukchi, Greenlandic and Bequia hunts, the issue surrounding the Makah was significantly more complex. This stemmed from the fact that the Makah ceased their gray whale hunts voluntarily in 1926 due to the dramatically overexploited gray whale stocks. Even though they merely took a very small number per year, they nevertheless considered any whale that was taken one too many. Since the gray whale population recovered only very slowly, the Makah did not conduct their traditional, ceremonial and subsistence-based whale hunts for more than 70 years. By the mid 1990s, however, gray whale stocks had recovered and the Makah wished to resume their whale hunts. Even though both the moratorium and the US Marine Mammal Protection Act (MMPA) of 1972 were then in place—we turn to the MMPA in the following section—they also had a treaty-based right to conduct whaling. Consequently, they agreed with the US government on support for non-commercial gray whale hunts and the US submitted a Needs Statement in 1996.

The United States is one of the most powerful actors in the IWC—and in world politics in general—and holds a strong record of whale conservation. Still, the Needs Statement for the Makah was not accepted immediately. Countries like Australia or Oman strongly opposed a resumption of Makah whaling. While Australia has in general supported aboriginal rights in the IWC, it questioned the traditionality of Makah whaling since the older generation of whalers that held the knowledge on traditional whaling had been gradually replaced by younger whalers who, according to Australia, did not hold such knowledge anymore. Oman, itself an outspokenly anti-whaling country, questioned the legitimacy of Makah whaling and the legitimacy of resuming it at the end of the twentieth century. After all, the Makah had not been whaling for 70 years and survived, so why would it be impossible to continue surviving without whaling? Also other countries questioned the Makah's whaling tradition in light of the 70-year break. In the end, however, in 1997 the Schedule was amended in a way that included the Makah take of the North Pacific gray whale stock, thus the US-presented Needs Statement was accepted.

The case of the Makah shows that the fact that an aboriginal people has a whaling tradition does not automatically provide reason enough for the IWC to accept this tradition and allocate an ASW quota. As Australia expressed in the deliberations surrounding Makah whaling, it rather hoped for a long-term phasing out of whaling than a resumption of it. In other words, even though the Makah have a legal right to conduct whaling, the overall wish to phase out whaling activities worldwide may constitute a normative obstacle to this right. Moreover, just because an aboriginal people has a history of whaling does not mean that its national government presents a Needs Statement in the first place and thus opens the door for an allocation of a quota. The case of the Ainu in Japan stands exemplary in this regard. As we have seen in Chap. 3, apart from the coastal whaling tradition in Japan, also the indigenous Ainu have hunted whales and seals since time immemorial.

Several issues coalesce in the question of why the Ainu, despite their whaling tradition, have not been recognised as ASW under the IWC. First and foremost, domestic politics play a role. The Ainu have struggled to be recognised as an indigenous people for decades and only in 1997 their status as an indigenous minority was accepted. Only since 2009, however, the Ainu's status within Japan is officially that of an indigenous people, after Japan has signed the UN Declaration on the Rights of Indigenous Peoples (UNDRIP) of 2007.[17] Since then, however, the relationship between the Ainu and the Japanese government has not improved, but has been marked by tensions. Negotiations between the Ainu and the Fisheries Agency concerning Ainu whaling have thus far been unsuccessful and therefore a Needs Statement from Japan concerning Ainu whaling has never been submitted. With Japan's withdrawal from the IWC, Ainu whaling might come into being, based on domestic law. Notably, throughout the IWC's history—as reflected in the Commission's annual reports—there has never been any reference to the Ainu. This means that a discourse on Ainu whaling, or even Ainu ASW, does not exist in the Commission. The Japanese government has thus not chosen to use this avenue to reopen whaling for human consumption in its domestic waters.

Instead, Japan has tried to introduce a different category, which essentially aligns itself with the 'needs' associated with ASW: Small-Type Coastal Whaling (STCW). After all, some coastal communities have a long-standing history of whaling, the most well-known of which is the dolphin and pilot whale drive in the town of Taiji in Eastern Japan. Before the moratorium, Taiji whalers also set out to hunt baleen whales, particularly minke, that were passing through the coastal waters of Japan. The Taiji whale hunt dates back several hundred years and has played an important socio-economic and cultural role in the community and in the region. After the moratorium, Taiji and other coastal whalers no longer had the chance to go after large whales despite their long-standing tradition and, arguably, their need for whale meat. In order for these communities to continue whaling, the Japanese government has since the late 1980s continuously argued for a whaling category between commercial whaling and ASW. While essentially the same line of argumentation as for aboriginal whaling, STCW naturally did not include an ethnic dimension, but

[17]United Nations Declaration on the Rights of Indigenous Peoples, 107th Plenary Meeting, 13 Sept 2007.

Japan argued along historical and needs-based lines. In 1987, for example, a report was presented to the Commission in which anthropologists and social scientists from Japan and elsewhere outlined the history and necessity of whaling for the communities in question.[18] While many Commission members were sympathetic towards the coastal communities, they were nevertheless suspicious towards Japan's motivations—a suspicion that has continued until Japan's withdrawal from the IWC. After all, an introduction of the STCW category would mean the allocation of a limited quota for baleen and sperm whales for the communities concerned. This, in turn, would mean a *de facto* lifting of the moratorium, which, for more than three-quarters of the Commission would be unacceptable. The fear of other countries following suit, paired with the fear of potentially ever-increasing quotas made the Japanese attempts fall on rather deaf ears. The last attempt occurred at the Commission meeting in 2018 when Japan's proposal for co-existence included the allocation of a quota for coastal whaling communities. As is known, the attempt failed and Japan announced its withdrawal shortly after the meeting.

The question of STCW and ASW go closely hand in hand. While ASW is by and large accepted by Commission members based on the historicity and necessity of whaling for aboriginal communities, the same does not account for non-aboriginal communities, for instance in Japan. The denotation of *aboriginal* is therefore crucial in this regard. But even though ASW is enshrined in the Schedule and in the working procedure of the Commission, it is by no means a given. The testament of this could be observed at aforementioned Commission meeting in 2018, when new ASW quotas were decided upon through a respective Schedule amendment. The specialty of this vote rested in the fact that the 6-year quota allocation would be extended to a 7-year allocation based on automatic renewal unless the Scientific Committee had concerns over the conservation implications of this automatic renewal. In addition to this new mechanism, the amendment also saw the inclusion of carry-over provisions, which would mean that quotas from a year that were not fully used could be used in the following year.

The discussions that arose were preceded by emotional testimonies of aboriginal whalers from Russia, Greenland, and the US (both Eskimo and Makah). Also the Commissioner of St Vincent & the Grenadines gave an emotional account of the benefits of the new system. While the overall majority of Commission members approved of the new system and thus reached the three-quarter majority to amend the Schedule, it was particularly Latin American states that voted against it. The automatic renewal was particular cause for concern and the states feared that this would mean either *de facto* commercial whaling, just conducted by aboriginal peoples, and might cause conservation issues in the future. Whether or not these concerns are justified remains to be seen. Either way, it is clear that even though ASW is an integral part of the IWC and legally possible, it is not to be taken for granted. After the vote in the Commission, a sense of relief grappled aboriginal delegations, as the image below demonstrates (Fig. 5.2).

[18]Freeman and Akimichi (1988).

Fig. 5.2 Relieved Alaskan delegates after the ASW vote at IWC67 © Nikolas Sellheim

5.3.7 Whaling and US Law: International Dimensions of Domestic Law

Although but one actor of the IWC, the US has developed powerful means to substantiate its long-standing anti-whaling policies both domestically and internationally. In fact, US domestic law relating to marine mammals and particularly to whales has been designed in a way in that it has an important external, international dimension.

The foundation for this dimension can be found in the 1970s and is a combination of several laws relating to fisheries conservation. The most important of these instruments is the so-called Pelly-Amendment to the Fishermen's Protective Act of 1967, which was adopted in 1971. The Pelly Amendment aimed at the imposition of trade barriers relating to marine products for those countries that undermine the effectiveness of an international fisheries conservation regime. Although initially aiming only at European countries that cooperated little in the protection of salmon stocks under the Northwest Atlantic Fisheries Convention,[19] with the overall discourse on whaling and the increasing calls for a moratorium on commercial whaling, the Amendment was also to cover the Whaling Convention. The Pelly Amendment allowed the Secretary of Commerce or the Interior to notify the US President of a

[19]International Convention for the Northwest Atlantic Fisheries, 8 Feb 1949 (157 UNTS 158).

country's breach of a fisheries protection agreement. The President, in turn, would then be able to impose seafood trade bans onto that respective country.

In the 1970s it was particularly Japan and the Soviet Union who were notified as exceeding IWC quotas for fin, sei and sperm whales. Since the US threatened these countries with trade sanctions, both accepted (and arguably also adhered) to the quotas set by the Commission. Also other countries that, at that time, were not members of the IWC received such notification and were threatened with sanctions. These countries, however, finally opted to join the IWC and to adhere to the quotas. Problematic in this account is that IWC rules and the Pelly Amendment do not necessarily align. While the Pelly Amendment is designed to support conservation agreements, it falls short of recognising the organisational set up of the respective agreement itself. In the case of Japan and the Soviet Union, both countries formally objected to the quotas and were thus not bound to them. The same accounted for Iceland when it was notified under the Pelly Amendment in 2011 as diminishing the effectiveness of the IWC by conducting commercial whaling. When Iceland rejoined the IWC in 2002, it formally, and legally, issued a reservation to the moratorium on commercial whaling and is therefore not bound to it. The Pelly Amendment neglects these internal procedures and inevitably places US views on how the IWC (or any other agreement is to function) onto other states.

A similar case has occurred in 2014 with respect to Iceland's trade in whale products. This time, however, it was not the IWC which was cause for concern, but the Convention on International Trade in Endangered Species (CITES) of 1973. To recall, CITES contains three Appendices. All species listed on Appendix I are fully protected and trade in them is not allowed. Amongst others, fin, sperm and humpback whales are listed on this Appendix. According to CITES rules, countries can lodge reservations to the listing and can conduct trade with non-members or other parties with reservations. Since trade in fin whale meat occurs mostly between Iceland and Japan, the latter of which also having also lodged a reservation, trade is therefore fully in conformity with CITES rules. A certification under the Pelly Amendment referring to undermining the effectiveness of CITES consequently negates the mechanisms within CITES itself.

The second important act in the United States to understand the external dimension of US law is the Marine Mammal Protection Act (MMPA), which the US Senate adopted in 1972. The MMPA is a far-reaching act which protects all marine mammals, irrespective of their conservation status, from commercial exploitation. Since the US has long recognised the close interaction of its indigenous populations with marine mammals, however, these are exempted from the provisions of the act, provided they do not conduct marine mammal hunts in a wasteful manner. In practice this means that marine mammal products cannot be sold on the US market and cannot be imported or exported, even for personal use. This furthermore means that despite the provisions of the ICRW and potential commercial whaling that is legal under the Convention, in US waters no whaling activities are permitted. In 1976 the Magnuson-Stevens Fishery Conservation and Management Act was passed whose main purpose was to prevent overfishing in US waters. Both the MMPA and the ICRW are explicitly listed as regimes protecting fisheries. If, therefore, a country is decimating whale populations that also occur in US waters, the Magnuson-Stevens Act allows for a certification to Congress, which in turn enables the Secretary of Commerce or the Interior to initiate the process under the Pelly Amendment.

Under the Magnuson-Stevens Fishery Conservation and Management Act, the Packwood Magnuson Amendment of 1979 is noteworthy. The amendment considers a reduction of 50% of a foreign country's fishing rights in US waters if it is in direct violation of the ICRW. The Secretary of Commerce certifies a country under this amendment.

So far, the US has not factually gone through with the imposing of trade barriers based on the certification under the Pelly or the Packwood Magnuson Amendment. But a certification corresponds rather to a yellow card in football, imposing pressure onto other states to change their behaviour. It is not a red card in the sense as the US would sanction other states. After all, since whaling states, such as Japan or Iceland, are close partners in many different areas of cooperation, trade sanctions would cause disruptions in the US-led free trade agenda. However, the existence of these amendments and the possibility of certification under both is likely to have contributed to Japan having withdrawn its objection to the moratorium.

5.4 Other International Instruments and Bodies

As we have seen in Chap. 4, the Convention on Migratory Species of Wild Animals (CMS) constitutes an important international instrument relevant for the protection of whales. Since the CMS is a framework convention that sets standards for the protection of migratory species, it is up to its parties to enter into agreements that more precisely apply the convention's provisions. To recapitulate, two mechanisms are possible in this regard: first, the conclusion of new, legally-binding agreements that enact the CMS for particular species and/or regions; second, the entering into Memoranda of Understanding, which are legally non-binding agreements, but essentially policy alignments and closer cooperation on a specific conservation matter between those states having entered into an MoU.

One agreement and one MoU have been concluded that serve as conservation tools for whales. One is the Agreement on the Conservation of Cetaceans in the Black Sea, Mediterranean Sea and contiguous Atlantic area (ACCOBAMS),[20] which was concluded in 1996 and entered into force in 2001. The ACCOBAMS is similar to its sister, the Agreement on the Conservation of Small Cetaceans of the Baltic and North Seas (ASCOBANS), see below, but differs in so far as it encompasses all cetaceans and not just small cetaceans. However, the main purpose of the ACCOBAMS is to achieve or maintain a favourable conservation status of all cetacean species and to prevent the deliberate taking of cetaceans. Moreover, the

[20] Agreement on the Conservation of Cetaceans in the Black Sea, Mediterranean Sea and contiguous Atlantic area.

parties agree to establish a network of specially protected areas for cetaceans. All littoral states to the Mediterranean Sea and the Black Sea are parties to the ACCOBAMS, so the agreement area encompasses both seas in total. In 2010, however, the 24 parties to the ACCOBAMS decided to include also the Spanish and Portuguese EEZs into the agreement's geographical coverage.

ACCOBAMS has focused on several 'traditional' conservation issues, such as population estimation, the reduction of ship strikes, fisheries-cetaceans interactions and associated recommendations for gillnet fisheries, or anthropogenic noise. Additionally, ACCOBAMS has also included matters related to cetaceans in captivity, for example by having provided guidelines for the release of cetaceans into the wild. Also public awareness, whale watching and issues related to whale strandings are well within the functionary scope of the ACCOBAMS. The further establishment and maintenance of protected areas for whales constitute a primary focus of the agreement, which are being identified based on areas of interest for marine mammals and those particularly affected by anthropogenic threats.

ACCOBAMS parties, however, have taken further multilateral steps to ensure the protection of marine mammals in their respective EEZs. The Agreement for the Creation of a Sanctuary for Marine Mammals (Pelagos Agreement),[21] for instance, creates a sanctuary in the Corso-Liguro-Provençal Basin between France, Monaco and Italy and essentially mirrors the provisions of the ACCOBAMS. However, it goes beyond the ACCOBAMS since it includes all marine mammals in the area and not merely cetaceans. Given the similarity with the ACCOBAMS, however, it is necessary to mention in line with it.

A legally non-binding MoU was concluded between several Pacific island states, Australia and New Zealand in 2006. The Memorandum of Understanding for the Conservation of Cetaceans and their Habitats in the Pacific Islands Region and Action Plan[22] focuses on the CMS-typical cetacean-related elements that also characterise the ACCOBAMS, such as threat reduction; habitat protection; research and monitoring; education and public awareness; Strandings and cetacean-based tourism. A special element of the South Pacific MoU is its Whale and Dolphin Action Plan, which includes the focus on cultural elements of whales and dolphins for the people in the region. Even though none of the small island states have even been whaling nations, twentieth century whaling has left significant marks on the abundance of whales. The MoU thus supports domestic education programmes that highlight the cultural importance of whales throughout the MoU's signatory states.

[21]Agreement for the Creation of a Sanctuary for Marine Mammals; original in French Accord Relatif à la Création en Méditerranée d'un Sanctuaire pour les Mammifères Marins, 25 Nov 1999.

[22]Memorandum of Understanding for the Conservation of Cetaceans and their Habitats in the Pacific Islands Region and Action Plan, UNEP/CMS/PIC-1/Inf/3, 15 Sept 2006.

5.5 Small Cetaceans

Small Cetacean Conservation

- No global conservation regime
- Regional multilateral conservation efforts, particularly under the Convention on Migratory Species of Wild Animals
- National initiatives, such EEZ-based marine protected areas

Contrary to the 'great whales' that are protected and the hunt of which is regulated by the International Whaling Commission, this does not account for small cetaceans—dolphins and porpoises (toothed whales). For decades, attempts to place small cetaceans under the wings of the IWC have failed. Instead, the international legal regime for their protection does not rest primarily on one organisation, but is a patchwork of national and international protective measures, some of which are also of relevance for the protection of large whales.

Broadly speaking, the protection of small cetaceans is primarily based on the initiatives of nation states themselves as the IWC does not provide for the management of small cetaceans. Over the years, the issue of small cetaceans has arisen frequently on the IWC's agenda, but many countries, especially those favouring the sustainable utilisation of whales, have opposed a broadening of the IWC's mandate. Still, the IWC's Scientific Committee has established a Small Cetaceans Sub-Committee in 1979, providing scientific advice for countries with threatened small cetacean species. In fact, a large number of resolutions has been passed that address the IWC's role in the conservation of small cetaceans, all of which have been consolidated into specific terms of reference for the Small Cetaceans Sub-Committee in 2014. Without directly expanding its mandate of the IWC towards small cetaceans, the IWC encourages its member states to conduct scientific work on small cetaceans and to contribute technically and financially to small cetacean conservation efforts.

The history of the interaction between small cetaceans and humans has not been marked by such a long-standing controversy as that of larger whales and humans. When the anti-whaling movement kicked off in the 1960s/70s, focus rested primarily on large (baleen) whales while small cetaceans were considered primarily as part of the greater 'Save the Whales' narrative. Denotations as 'the dolphin' or 'the whale' neglected the inherent differences in species composition and stock statuses that characterise all cetacean species. In the past, primary threats to small cetaceans did not stem from overhunting, given their relatively small size. Instead, other anthropogenic threats, which are equally affecting large whales, severely impact small cetaceans: habitat degradation or loss, noise and other pollution, ship strikes, climate change or by-catch. Several small cetacean species are severely threatened. For instance, the vaquita (*Phocoena sinus*) has been dangerously depleted as result of being by-catch in the extensive totoaba (*Totoaba macdonaldi*) fisheries in the Baja California. At the point of writing, less than 30 animals are expected to exist. Also the Baiji, or Yangtse river dolphin (*Lipotes vexillifer*) in China has been dramatically reduced—to a degree that it was declared functionally extinct in 2007 since a survey was not able to locate any individual. Another example is the Ganges river dolphin (*Platanista gangetica ssp. gangetica*), whose range states India, Bangladesh and Nepal have placed it under strict protection due to its decreasing conservation status.

In order to protect small cetaceans from further habitat loss, nation states have either put in place species-specific conservation regimes or marine protected areas, which include small cetaceans. In countries such as the Faroe Islands or Japan, where different dolphin and pilot whale species are hunted for their meat or for sale to aquariums, strict regulatory, science-based measures are in place that aim to ensure a sustainably favourable conservation status of these species. In order to achieve this, cooperation is necessary. This is especially the case since the conservation status of many small cetacean species is data deficient, which means that international authorities, such as the IUCN, do not have enough data available to determine whether the population trends of specific species are increasing or decreasing. This is particularly the case for the various beaked whale species (*Ziphiidae*), as Erich Hoyt demonstrates.[23]

Since the focus of this book is not to delve into national protection strategies, but to consider the international legal environment, let us now turn to those international instruments, which have specifically been designed to protect small cetaceans. A key regime in this regard is the CMS, which was introduced in Chap. 4. Since the CMS is a framework convention and therefore sets standards for other means of international cooperation in the context of migratory species, it is particularly European countries that have come together to adopt a regime for the protection of small cetaceans. This regime, the Agreement on the Conservation of Small Cetaceans of the Baltic and North Seas (ASCOBANS),[24] was concluded between several states surrounding the Baltic and North Seas. In 2008, ASCOBANS' name was changed to Agreement on the Conservation of Small Cetaceans of the Baltic, North East Atlantic, Irish and North Seas, thus extending its area significantly, now including the Atlantic waters of Spain and Portugal, both of which, however, are not party to the Agreement (see Fig. 5.3).

The history of ASCOBANS goes back to the mid-1980s when cetacean scientists suggested the inclusion of small cetaceans onto Appendix II of the CMS since many species in the Baltic and North Seas had declined dramatically due to habitat deterioration, by-catch and other forms of disturbance. At the second Conference of the Parties of the CMS in 1988, several small cetacean species were thus added to Appendix II, which caused Norway to file an official objection to this listing. Germany, on the other hand, proposed a global agreement for small cetaceans, but the idea was not further pursued. After all, controversy over the whaling moratorium in the IWC did not bode well for a global agreement on small cetaceans.

[23]Hoyt (2011), p. 46.

[24]Agreement on the Conservation of Small Cetaceans of the Baltic and North Seas, 17 March 1992 (1772 UNTS 217).

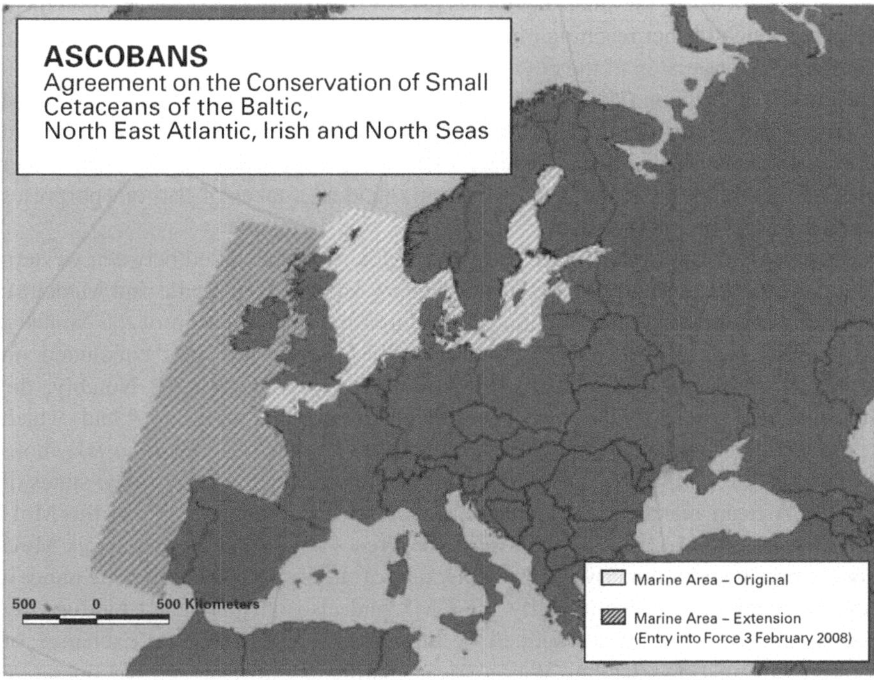

Fig. 5.3 Geographical scope of ASCOBANS. © Map courtesy of ASCOBANS (ascobans.org)

This notwithstanding, it became clear that some sort of agreement for the conservation of small cetaceans would have to be found. Particularly the Netherlands fostered the idea of a regional agreement for the North and Baltic Seas in order to achieve quick results. At the initiative of Sweden, an agreement text was drafted and finally, in 1990, a first Memorandum of Understanding on Small Cetaceans in the North Sea was signed in The Hague. Since an MoU is not legally-binding, it lacked 'teeth' for the effective protection of small cetaceans. Sweden, therefore, took up the initiative and developed the MoU further for it to turn into a legally-binding agreement for the states concerned. Thus, in 1992 ASCOBANS was opened for signature.

ASCOBANS aims to provide for a favourable conservation status for all small cetaceans, i.e. all toothed whales except for sperm whales, in the treaty area. This means essentially two things: first, even though the agreement does not spell it out, the lethal taking of small cetaceans is prohibited. This is true particularly in light of the Convention on the Conservation of European Wildlife and Habitats, which strictly prohibits the taking of cetaceans, as we have seen in Chap. 4. Second, parties to the agreement are to reduce those factors that contribute to the decline of small cetaceans in the treaty area. This pledge translates into the reduction of noise and other pollution, the modification of fishing gear to reduce by-catch and to reduce the impact on those species that serve as food sources for small cetaceans in the treaty area. While focusing on small cetaceans, ASCOBANS is therefore an agreement which has much further reaching impacts, complementing the work of other regional bodies for the protection of the environment, such as the Baltic Marine Environment Protection Commission (HELCOM). Not surprisingly, ASCOBANS and HELCOM cooperate rather closely on issues related to small cetaceans and the reduction of threats associated with them. Primary efforts that have been undertaken under ASCOBANS relate to abundance estimation, incidental takes of harbour porpoises and to the reduction of by-catch.

Under the CMS, a small cetacean-specific MoU was concluded between western African states and Macaronesia (Azores, Canary Islands, Cap Verde and Madeira). The Memorandum of Understanding Concerning the Conservation of the Manatee and Small Cetaceans of Western Africa and Macaronesia[25] was concluded on 8 October 2008 and encompasses 16 African states and Portugal. Notably, the MoU was also signed by the three land-locked states Mali, Niger and Chad, which is particularly important for the African manatee (*Trichechus senegalensis*) whose habitat stretches across the waterways of these countries. Why other range states of both the African manatee and small cetaceans have thus far not signed the MoU cannot be ascertained. Be that as it may, the West African marine mammals MoU strives directly and indirectly for the protection of these species. On the one hand, it encourages the signatories to ratify the CMS and also other relevant biodiversity treaties. On the other, threat reduction, habitat restoration and the exchange of scientific information related to West African marine mammals are core cooperative features of the MoU.

According to the Action Plan, which is an integral part of the MoU, the cultural aspects of marine mammals are to be considered while the non-lethal sustainable use, meaning marine mammals for income generation without their killing, is a fundamental element of the Action Plan. The interrelation with other international agreements stands out in the Action Plan since the MoU's area of application corresponds to the area of the 1981 Convention for Co-operation in the Protection and Development of the Marine and Coastal Environment of the West and Central African Region (Abidjan Convention).[26] Although the Abidjan Convention is not directly applicable to marine mammals, it nevertheless deals with threats to the coastal regions and marine areas of the parties. Consequently, it is also of importance for small cetacean and manatee habitat restoration and threat reduction.

---

[25]Memorandum of Understanding Concerning the Conservation of the Manatee and Small Cetaceans of Western Africa and Macaronesia, 8 Oct 2008.

[26]Convention for Co-operation in the Protection and Development of the Marine and Coastal Environment of the West and Central African Region, 23 Mar 1981 (20 ILM 746).

5.5.1 Narwhal and Beluga Protection in Eastern Canada and Greenland

Two of the most iconic and impressive whale species are arguably the *Monodontidae*: the 'white whale', beluga (*Delphinapterus leucas*) and the narwhal (*Monodon monoceros*). These primarily (sub-)Arctic species have been subjected to long-standing subsistence hunts by Canadian and Greenlandic Inuit for centuries. The long-term effects of these hunts is unknown. However, in combination with commercial exploitation of these species, towards the end of the nineteenth century *Monodontidae* populations had declined drastically. Particularly the pursuit of ivory of the narwhal, whose single tooth provides it with its charismatic appearance, resulted in the species being drastically overhunted. The Canadian government enacted its first regulations of beluga management in 1949, followed by narwhal regulations in 1971. Both regimes considered these whales under an ecosystem approach and provided for the protection of narwhal and beluga habitat, relied on scientific advice, and regulated management and harvests. The federal dominance of beluga and narwhal management was challenged with the adoption of the Canadian land claims agreements: the 1975 James Bay and Northern Quebec Agreement; the 1984 Inuvialuit Final Agreement; the 1993 Nunavut Land Claims Agreement; and finally the 2005 Labrador Land Claims Agreement. Generally speaking, these agreements provide aboriginal resource users with the right to continue using them even in light of federal regulations that say otherwise. Moreover, aboriginal peoples in Canada have a constitutional right to exploit fish resources to the extent they deem necessary.

Against this backdrop, an aboriginal appellant challenged the Department of Fisheries and Oceans (DFO) after he had been arrested in 1984 for using wrong fishing gear. In 1988 the so-called 'Sparrow case' was launched in which the constitutional, collective rights of aboriginal peoples for their own *fish* resources was brought forth. In 1990 the Supreme Court of Canada ruled that the Canadian government can only restrict and limit aboriginal fish resource use if there is proof that aboriginal harvests threaten the conservation status of the resource. Since under Canadian law, the definition of 'fish' includes also marine mammals, the ruling had tremendous impact on the way belugas and narwhals were to be managed in the eastern Canadian Arctic.

Mindful of the then ongoing Sparrow case, and taking into account that belugas and narwhals are exploited by both eastern Canadian Inuit and Western Greenlanders, a Memorandum of Understanding between Canada and Greenland was signed in December 1989 for the management of belugas and narwhals migrating between the two countries. Under the MoU, which aims for the rational management, conservation and optimum utilisation of living resources of the sea—as per the UNCLOS—a Joint Commission was established in 1991 which consists of two Commissioners for each country: one representing the government, the other representing the hunting communities. While the main work of the Joint Commission is carried out by its Scientific Working Group, comprising of fisheries scientists and whale biologists, thus putting great emphasis on 'western' science, the work is also to be supplemented by Inuit traditional knowledge. To this end, the current set up of the Joint Commission also includes a Traditional Knowledge Working Group as well as a User-to-User Working Group. The approach of the Joint Commission is therefore not to take top-down decisions that might negatively impact resource users, but rather to include the concerns and knowledge(s) present in hunting communities.

While the Commission consists of government officials, it does not have the mandate to make binding decisions. Instead, it advises the respective government bodies responsible for the management of the species in question. In how far these recommendations are adhered to is questionable. For instance, while in 2008 the Joint Commission recommended a reduction of hunting quotas by Greenland due to the drastic reduction of beluga and narwhal stocks, Greenland nevertheless increased its quota. The Greenlandic government justified this step by referring to the concerns of Greenlandic hunters who considered a reduced quota as too restrictive. Moreover, relying on the observations of hunters in the area, Greenland was convinced that narwhal and beluga populations were stable enough to sustain higher quotas.[27] The conflict between scientific assessments and traditional knowledge become apparent in this context. Already in 1998 Freeman et al. noted that one of the challenges of the Joint Commission was to meaningfully integrate traditional knowledge into its decision-making procedures.[28]

5.5.2 Inuvialuit-Iñupiat Beluga Whale Management

The Northwest Territories in Canada and the North Slope of Alaska is the home region of the Inuvialuit and the Iñupiat respectively. These two Inuit groups have lived and hunted in the region for several hundred years, including whales, seals and polar bears. In 1988, the two groups entered into an agreement on polar bear management in the Southern Beaufort Sea (see Sect. 7.4), which was designed to be a user-to-user agreement, concerned with the interests of the hunting communities as well as with the conservation of polar bears. Another important species in the region is the beluga whale. Since belugas traverse national boundaries, both the Inuvialuit Game Council and the North Slope Borough considered it necessary to establish a similar agreement to the polar bear agreement for the beluga.

In 1988, therefore, the Alaska and Inuvialuit Beluga Whale Committee (AIBWC) was established. One primary reason for its establishment was the need for the collection of catch data and the need for more biological information on the beluga in Alaska and the Western Canadian Arctic. Especially information on ice entrapments, ship strikes and pollution. What made the AIBWC rather unique in its set up is the fact that half of its members comprised hunters organisations from all over the region. The other half consisted of government representatives, scientists and technical advisors. This setting therefore allowed the AIBWC to conduct broad

[27] Fitzmaurice (2015), pp. 298–300.

[28] Freeman et al. (1998), p. 143.

and inclusive studies on belugas in the region while it also provided for the development of management plans. In 1994, however, the AIBWC was changed in both scope and structure. From then on it was named the Alaska Beluga Whale Committee (ABWC), indicating that the cooperation with the Canadian side no longer existed. Whether that was the case I am not able to confirm with full confidence though. What is clear is that by the year 2000 an agreement between the Inuvialuit of Canada and the Iñupiat of Alaska was concluded: the Inuvialuit Inupiat Beaufort Sea Beluga Whale Agreement.[29] While in principle picking up on where the AIBWC had left off, the agreement established the Inuvialuit-Iñupiat Beluga Whale Commission—an extremely low-profile commission consisting four Inuvialuit hunters representatives, three Iñupiat hunters representatives, and one representative of the Kivalina Whaling Captains Association in Alaska.

Based on the agreement, the purpose of the commission is to facilitate information exchange, particularly on the sex of the whales landed, the locations of their takes and the numbers that have been taken. The agreement itself is therefore not a legally binding agreement, nor is the commission a management body. Instead, it communicates the results of hunting activities on both sides of the border—which follow the respective national management plans—to each other.

No additional information on the commission appears to exist. Any further information is most welcome for subsequent editions of this book.

5.5.3 The Conservation of Dolphins in the Pacific Ocean

As is commonly known, tuna fisheries have one important drawback: many dolphins are taken as by-catch. This is particularly the case in the eastern tropical waters of the Pacific where large yellowfin tuna (*Thunnus albacares*) share the same habitat as several dolphin species. In fact, the bond between these two species is so strong that tuna fishers used to use dolphins as 'bait' to be able to fish tuna: when they rounded up dolphins that were accompanied by tunas, the latter followed the dolphins, enabling the fishers to take both tuna and dolphin at the same time. In the past, this has led to massive numbers of dolphin by-catch. It is estimated that in this manner more than 6 million dolphins have been taken. The public resentment over the by-catch of dolphins was a driving force behind the adoption of the US Marine Mammal Protection Act in 1972, which holds specific provisions on dolphins. Due to US efforts, by 1980 the annual dolphin by-catch had been reduced from around 500,000 animals to 20,000 animals per year.

Yet, since the United States is not the only country fishing tuna in the Pacific, already in 1950 the Inter-American Tropical Tuna Commission (IATTC) was established, comprising a large number of states from the Pacific Islands, North and South America, Japan, and the European Union. Although the US regulations proved to be successful, other states that did not have comparable regulations

[29]Inuvialuit Inupiat Beaufort Sea Beluga Whale Agreement, 3 Mar 2000.

expanded their fisheries, driving up dolphin mortality once again. Although the IATTC had started to put in place a dolphin conservation programme, in 1986 due to an international observer scheme it became clear that despite drops in dolphin mortalities, they had risen to around 133,000 per year again. By then, the United States, one of the main markets for Pacific tuna, put in place requirements for 'dolphin-safe' labels, which put pressure onto other tuna-fishing states. Moreover, in 1990 the US imposed an import embargo for tuna without the label, even accounting for tuna that could potentially have resulted in dolphin mortalities unless reliably demonstrated by the fishing nations that that was not the case. In other words, the US banned tuna imports unless it was shown that they were conducted in a dolphin-safe manner. In 1991 this led to Mexico and other states taking the embargo to the World Trade Organization (WTO) as being in breach of the free-trade provisions of the General Agreement on Tariffs and Trade (GATT).[30] However, without a final ruling of the WTO, Mexico and the US settled this dispute 'out of court'. One of the reasons for this settlement was the Agreement for the Conservation of Dolphins (La Jolla Agreement),[31] which was concluded between several—not all—member countries of the IATTC. The La Jolla Agreement aims to progressively reduce dolphin mortalities and to ultimately phase them out entirely. To achieve this, a Review Panel was established which set annual 'Dolphin Mortality Limits', starting at almost 20,000 in 1992 but dropping to less than 5000 by 1999.

While the La Jolla Agreement was an agreement that proved to be successful, it was nevertheless not legally binding while some of its signatories were not members of the IATTC. In 1995, therefore, the governments of the signatory states met in Panama and decided to conclude a legally binding convention which would require states to put in place measures to reduce dolphin mortalities. In 1998, the successor to the La Jolla Agreement was adopted, the Agreement on the International Dolphin Conservation Program (AIDCP),[32] entering into force in February 1999. The AIDCP aims to reduce dolphin mortalities to less than 5000 dolphins per year. The AIDCP has a membership of sixteen at the time of writing. While strictly tied to the IATTC, the AIDCP is a stand-alone agreement with its own membership. A key provision of the AIDCP is an observer scheme which requires tuna vessels to take international observers on board, who oversee the operations of the vessel *vis-à-vis* the convention's provisions. The members of the convention meet back-to-back with the annual meeting of the IATTC and establish dolphin mortality caps on a yearly basis and for each dolphin stock, yet not exceeding 5000 per year.

[30]General Agreement on Tariffs and Trade, 14 Apr 1994 (55 UNTS 194).

[31]Agreement for the Conservation of Dolphins, 21 Apr 1992.

[32]Agreement on the International Dolphin Conservation Program, 21 May 1998.

5.6 Summary and Conclusion

International law that has been designed specifically for whales is manifested in the three agreements on whaling. The first Convention on the Regulation of Whaling was concluded already in 1931 and was the first agreement that dealt exclusively with the hunt for whales. It was concluded at a time when commercial whaling was at full swing and states such as Norway, the UK or the United States were heavily engaged in producing whale oil for the international market. Since at that time whale populations had already been significantly depleted, the primary whaling states agreed that an international agreement for the regulation of whaling is a necessity in order for the whaling industry to continue operating profitably.

Even though the 1931 convention was rather revolutionary at that time, it did not meet the expectations of the whaling parties. One reason was that it was also open for states that had no immediate interest in whaling and which would undermine the conservation efforts put forward by states that were in fact whaling. On the other, it became clear that the regime was not able to deal with the special situation in the Antarctic whale hunts, particularly concerning areas such as South Georgia. Therefore, under the leadership of the UK a second conference was convened in 1937 to conclude the International Agreement on the Regulation of Whaling. The London Agreement was only signed by a small number of states and with the outbreak of the Second World War, regulatory efforts concerning whaling were made void. After all, whales provided a valuable source of oil, which was needed for the war machinery. Both the 1931 and 1937 regimes were neither signed or ratified by Japan, which was an emerging commercial whaling nation at that time.

After the war, under the leadership of the United States, the issue of an international whaling regime was once again picked up and in November and December 1946 a conference, which included Japan, was convened in Washington, DC. While the United States pressed for more conservation, other states, such as Japan, the Netherlands or the traditional whaling states Norway and UK, aimed for sustainable utilisation in order to ensure a thriving whaling industry. Under these premises the International Whaling Commission was established. Over time, perception on whaling, particularly in Western states, changed and public opinion increasingly called for a cessation on commercial whaling. The United States was the first government to impose a ban on all commercial marine mammal hunts through its Marine Mammal Protection Act in 1972. But also on an international level, at the 1972 Stockholm Conference in the Human Environment, the call for a global moratorium on commercial whaling was uttered.

In 1982 the IWC decided on a zero-catch quota for all whales which it has the competence to regulate, to come into effect in the Antarctic whaling season 1985/86. This so-called moratorium was to be temporary and subject to a review in 1990. It must be emphasised that the hunt for toothed whales, except for the sperm whale, is not affected by the moratorium since the IWC does not have competence to manage small cetaceans. The Taiji dolphin drive or the Faroese pilot whale drive are thus not internationally banned. Be that as it may, in order to make changes to the Schedule of the Convention, which sets quotas, a three-quarter majority is necessary. Since at the time of the moratorium and thereafter the dominance of Commission members had shifted towards a no-whaling agenda, pro-whaling states such as Japan, Norway or Iceland have so far failed to garner the necessary majority for a Schedule amendment and lift the moratorium. Although different avenues have been tried, the rift between pro- and anti-whaling states in the IWC could not be overcome. This prompted Japan, one of the main financial contributors to the Scientific Committee, to announce its withdrawal in December 2018.

Although a moratorium on commercial whaling is in place, this does not mean that the IWC does not grant any whaling at all. As per the whaling convention and it Schedule, whaling for scientific purposes and Aboriginal Subsistence Whaling are permitted. The former had been conducted by Iceland and particularly Japan and has often been labelled as a guise for commercial whaling, as a means to circumvent the moratorium and thus as a loophole that these states use. The International Court of Justice ruled, at the application of Australia, in 2014 that the then ongoing Japanese whaling programme was not for the purposes of scientific research, and ordered Japan to cease these operations. The ICJ did *not* rule that scientific whaling in Antarctic is in contravention of the convention, which prompted Japan to readjust its Antarctic scientific whaling programme to the criteria set forth in the Judgement. With Japan's withdrawal from the IWC it announced that it would cease its Antarctic whaling programme entirely. Instead, scientific and commercial whaling would from now on only occur in Japanese waters and in the North Pacific, the latter of which will be phased out by 2020 as well.

In terms of ASW, even though it is an integral part of the current whaling regime, it is not a given. The Makah, for instance, had to struggle rather hard to get the Commission to allocate a small gray whale quota. Also for other aboriginal communities, a new quota is always a reason to celebrate since the IWC does not uniformly support the fact that in the twenty-first century some communities are still dependent on whales.

Indeed, although the United States are a strong supporter of aboriginal whaling, it has legislation in place that allows for the imposition of trade sanctions on other states if they violate the effectiveness of the whaling convention or other regimes, such as CITES, that aim to protect whales. This domestic legislation thus has an external dimension which is geared to put pressure on states to change their behaviour. Up to the point of writing in the fall of 2019, no trade sanction has been imposed on certified states, but it is undeniable that this constitutes a tool to impose the US view on others that cannot be neglected.

With Japan's withdrawal, which came into effect on 1 July 2019, the International Whaling Commission and with it the global whaling regime stands at a crossroads. Based on article 65 of the UNCLOS, states are required to cooperate in the management and conservation of marine mammals through appropriate organisations. It is unclear what these 'appropriate organisations' in practice are, as we have explored in Chap. 4. Either way, it remains to be seen whether Japan will initiate the formation of a new organisation, maybe along with other Asian states, such as China or South Korea, or introduce matters relating to whaling also into other organisations. Also the future membership of the IWC is not entirely clear at this point. After all, other pro-sustainable use countries, particularly from the Caribbean, Africa or the Pacific may follow suit and leave the IWC altogether. After all, the most vocal and most powerful—political and economically—champion of the sustainable use of whales is no longer a member of the Commission. However, as a Japanese official mentioned to me, Japan supports other states championing the principle of sustainable use to remain in the IWC. Inevitably, however, the discourse within the IWC will shift towards an increasingly no-use, preservation agenda. This was underlined by the 2018 Florianópolis Declaration.

Contrary to large whales, small cetaceans do not enjoy the same degree of structured, international protection. Despite attempts to expand the IWC's mandate to include small cetaceans, all of these have thus far been unsuccessful. Indeed, the WWF calls small cetaceans 'The forgotten whales'[33] given that they do not enjoy such international attention or international regulatory framework for their protection. Instead, several regional agreements for their protection have sprung up in the Mediterranean and Black Seas, Western Africa, the South Pacific, and the North and Baltic Seas and in the Arctic. While some of these are legally binding, some are not and states cannot be held accountable in case they do not comply. The Convention on Migratory Species of Wild Animals serves as the umbrella convention for these agreements and is thus a key convention for international protection efforts related to small cetaceans.

Whaling is and will be the source of controversy. In how far the international whaling regime will develop is impossible to predict. But even though Japan has now the possibility, and the motivation, to resume commercial whaling, it will do so based on scientific data. It is doubtful that a new tragedy of the commons, or ruthless overexploitation of whales, will occur in the future. After all, also whaling states have learned from the mistakes of the past and given that the demand for whale products is limited worldwide, I remain hopeful that whales will be around for many generations to come.

Questions and Research Tasks

1. Determine whether there might be a way to overcome the divide within the International Whaling Commission.
2. Select one member of the International Whaling Commission and provide insight on why it joined the Commission and what its line of argumentation concerning whales and whaling are.
3. Determine whether an international agreement for all whale species would be a feasible option, what its benefits and disadvantages are, and who could be a party to it.
4. Discuss the advantages and disadvantages of a global agreement for small cetaceans.

[33] WWF (2009).

References

Bockstoce JR (2018) White Fox and Icy seas in the Western Arctic. The Fur Trade, transportation, and change in the early twentieth century. Yale University Press, New Haven

Dorsey K (2013) Whales & Nations. Environmental policy on the high seas. University of Washington Press, Seattle

Fitzmaurice M (2015) Whaling and international law. Cambridge University Press, Cambridge

Fitzmaurice M, Tamada D (eds) (2014) Whaling in the Antarctic. Significance and implications of the ICJ judgment. Brill Nijhoff, Leiden

Freeman MMR, Akimichi T (eds) (1988) Small-type coastal whaling in Japan. Report of an international workshop. Boreal Institute for Northern Studies, Edmonton

Freeman MMR, Bogoslovskaya L, Caulfield RA, Egede I, Krupnik II, Stevenson MG (1998) Inuit, whaling and sustainability. AltaMira Press, Waltnut Creek

Hoyt E (2011) Marine protected areas for whales, dolphins and porpoises, 2nd edn. Earthscan/ Routledge, Abingdon

ICR (2011) Questions and Answers. https://www.icrwhale.org/QandA1.html

IWC (1973) Twenty-third report of the commission. IWC, London

Miller AR N Dolšak (2007) Issue linkages in international environmental policy: the international whaling commission and Japanese development aid. Glob Env Pol 7(1):69–96

Morikawa J (2013) Whaling in Japan. Power, politics and diplomacy. Hurst & Company, London

Prime Minister of Japan and his Cabinet (2018) Statement by Chief Cabinet Secretary, December 26, 2018 https://japan.kantei.go.jp/98_abe/decisions/2018/_00008.html?fbclid=IwAR27iTMRPkNlSCrwEri9RPTbDzM6wN08NgHHnB7nWPHyYBuKq97dcLproms

Taniai M (2018) Intervention at IWC67. https://www.youtube.com/watch?v=jrzUAVEtcU4, at 01:16:45

WWF (2009) Small cetaceans. The forgotten whales. WWF, Gland

Further Reading

Epstein C (2008) The power of words in international relations. Birth of an anti-whaling discourse. The MIT Press, Cambridge

Friedheim RL (ed) (2001) Toward a sustainable whaling regime. University of Washington Press, Seattle

Gillespie A (2005) Whaling diplomacy. Defining issues in international environmental law. Edward Elgar, Cheltenham

Stoett PJ (1997) The international politics of whaling. UBC Press, Vancouver

The International Legal Framework for Seals

6

6.1 Introduction

Seals belong to the species of marine mammals with the longest history of legal regulation. Even though, at the time of writing in 2019, they do not play as prominent a role anymore as whales in international discourse, they have served as important conduits for legal change towards non-utilisation. By and large, the discourse on seals and whales shows significant overlaps concerning their perceived status of being endangered and a strong no-use narrative. A significant difference exists, however: there has never been the seal-equivalent to a 'superwhale'—a 'superseal'—that is social, intelligent and sings songs. Instead, the discourse on seals has been much less shaped by criteria that make it almost human-like, but instead has been shaped by emotional responses. Images of 'baby seals' that emphasise childlike looks and fragility and the 'weeping' mother after its 'baby' has been killed by commercial sealers have not missed its mark and have therefore burned themselves into (western) public consciousness. In the 1990s in Germany, where I grew up, it was not uncommon to find 'Save the Seal' stickers on cars—also driving through a Drive-Through of McDonald's, the irony of which cannot escape the heedful observer—and a seeming consensus on the ruthlessness of commercial sealing. After all, how can one kill such a cute baby seal and cause so much harm to the loving mother?

As has been demonstrated on countless occasions, reality is somewhat more complex and the legal regimes all over the world that deal with seals differ, depending on the traditionality and currentness of sealing. It is clear, however, that whenever seals and humans have co-existed, they have also interacted, mostly to the disadvantage of the seal. As we have seen in Chap. 2, this interaction has occurred all over the world. Seals have played, and still do, three distinct roles for humans: as a source of food; as a commercial commodity (for oil and fur); and as a nuisance. Moreover, seals also play a role as a tourist attraction in seal watching tours all over the world. Most prominent are probably the California sea lions (*Zalophus*

N. Sellheim, *International Marine Mammal Law*,
https://doi.org/10.1007/978-3-030-35268-4_6

Fig. 6.1 Tourists watching sea lions at Pier 39 in San Francisco © David Bauer

californianus) at Fisherman's Wharf in San Francisco, which serve as a tourist magnet (see Fig. 6.1).

From a legal perspective, 'the' seal has not really existed until recently, however. This recent development is manifested in the 2009 trade ban on all seal products, which was put in place by the European Union (EU)—a union of 28 European states which share a common market. We will get to know this ban later on in this chapter. Prior to this approach, which has also been mirrored in other regions of the world, as we will see, seals were legally considered on a species to species basis. Hence, while the EU and other states consider seals as a whole, traditionally it was individual, and regionally-bound, species that were subject to regulation. Since an international regulatory forum in the form of a 'sealing commission' has not taken form, it was countries in bi- or multilateral efforts that have taken steps to regulate sealing—first in the interest of humans, later in the interest of the seals themselves.

6.2 The Nineteenth Century

Apart from national legislation that has seen the regulation of hunting—and thus sealing—activities for centuries, two regions stand out that have seen extensive multilateral regulation of sealing: the North Atlantic and the North Pacific. While in the former there were scattered agreements, the latter was subject to a multi-decade regime that regulated the hunt for Northern fur seals (*Callorhinus ursinus*).

While the longest regime could be found in the North Pacific, it was the North Atlantic which yielded the first actual regime to regulate sealing: the Jan Mayen Fisheries Treaty was concluded between Germany, the Netherlands, Norway, the United Kingdom and Russia in 1875 and relates to the sealing activities at Jan Mayen, a small island in the Greenland Sea in the North Atlantic. At the time of the treaty's conclusion, it was unclear who had sovereignty over the island since, on the one hand, Dutch whalers maintained a base there while Norway claimed to have discovered it first. The issue was only resolved in 1921 when Norway opened a permanent meteorological station at Jan Mayen and claimed sovereignty in 1926, confirmed by the League of Nations in 1929. Be that as it may, already in the 1850s it became clear that the abundant harp seal (*Pagophilus groenlandicus*) herds were drastically overhunted with numbers up to 400,000 per year. In order to continue hunting the seals, the Jan Mayen Treaty did not provide effective protection as such, but rather provided a means for the countries involved to streamline their respective national legislation pertaining to sealing. It was therefore more of a common-day framework convention that each party implemented according to its own means. The main feature of this little-known treaty was that it established a clearly defined hunting area and hunting seasons. In how far these were efforts were successful and effectively protected the abundant seals cannot be ascertained, but the Jan Mayen treaty is important in so far as it constituted the first international marine mammal-related agreement that regulated hunting activities.

While the Jan Mayen Treaty in the North Atlantic was the first of its kind, the North Pacific, and in particular the seal hunts on and surrounding the Aleutian Pribilof Islands, were subject to long-standing regulation that resulted in a regulatory regime that would last for almost 100 years. The Pribilof Islands are two islands, St George and St Paul, which now belong to Alaska (United States), but which were discovered in the late 1780s by the Russian commander Gerasim Gavrilovich Pribilof. No native population could be found on these two islands at that time. The discovery of the islands proved to be a gold mine for the Russian fur traders, who had expanded their grounds of operation in the North Pacific in their search for sea otters and fur seals. Given that the Pribilof Islands were uninhabited, they constituted the main two breeding grounds for the Northern fur seal, providing Russian fur traders with a seemingly unlimited number of furs. Word spread quickly and soon also American, Japanese and British (Canadian) schooners were heavily engaged in the hunts for fur seals, causing an unregulated decimation of the abundant seal herds. Within around 10 years after the discovery of the Pribilofs, the Northern fur seal populations were close to collapse.

To better control the fur seal exploitation at the Pribilof Islands, in 1799 the *Russian-American Company*, a Russian-led company for the expansion of Russian hegemony in America was given control, thus *de facto* constituting a first regulatory effort. This effort, however, proved to be rather unsuccessful and the decimation of the seal herds continued unabated, even leading to around 100,000 furs having to be disposed of due to an overabundance on the international markets. This did not escape the Russian authorities and in 1806 Russia's Grand Chamberlain—the official to manage a royal endeavour—Nikolai Rezanov ordered a complete halt of seal hunting at the Pribilofs. This decree did not last longer than 1808, however, and until 1821 the seal hunts continued without significant efforts to regulate them.

That year saw the first regulation to control the seal hunts. The order saw a maximum take of 50,000 seals per season, a full halt of the hunts every 5 years, and the sparing of bulls and pups. One year later, specific areas in the Pribilof Islands were *de facto* declared sanctuaries for specific time periods. For St George for the years 1826/27 and for St Paul for the years 1822–1824 and 1835–1837. By 1847, also 'sacred cows' were to be spared from the kills, seal cows that played an important role for the longevity of the seal herd. This meant that the focus of the seal hunts had effectively been shifted to immature bachelor seals. Paired with the quota of 50,000 seals per season, by 1867, when Alaska was sold to the United States, the Northern fur seal herds at the Pribilof had recovered.

With the sale of Alaska, the US legal system and authorities took over control. As per the 1868 Alaska Customs Act it was the Department of Treasury that controlled the hunts of fur-bearing animals in Alaska. This was particularly relevant since in the 1868 season after Russian hegemony had made way for the United States, no enforceable control mechanism was in place, leading to massive seal hunts, far outnumbering Russian hunts in the year before. On St Paul alone, for instance, 250,000 fur seals, including sacred cows, bulls and pups, were killed, resulting in a full ban on the killing of all fur-bearing animals in Alaska. Additionally, the Fur Seal Service was established, overseeing the implementation of government policies regarding the seal hunt. This became particularly relevant when Congress in a Joint Resolution made the Pribilof Islands special government reserves, making it illegal for anybody to set foot on the islands.

Since the interest in fur seals remained unchanged, however, a total ban on sealing proved unrealistic. Therefore, on 1 July 1870, the Act to Prevent the Extermination of Fur-bearing Animals in Alaska was adopted. This act established a principle ban on sealing, but allowed the Secretary of the Treasury for the leasing of the right to hunt seals in June, July, September and October to 'responsible parties' for a period of 20 years. Following the Russian example prior to the sale of Alaska, the catch limit was set to 75,000 seals for St Paul and to 25,000 seals for St George per year. Apart from this limit, however, the first exemption for the native population of Alaska was inserted, which was still allowed to hunt seals and other fur-bearers for subsistence purposes. In 1870, only one company, the *Alaska Commercial Company* (ACC), was given the lease to hunt seals, which also adhered to the set provisions.

Even though regulatory efforts were in place and the ACC performed well in regard to its compliance, technological advances still put the Northern fur seal populations under increased pressure: larger ships that were able to hunt out at sea and the increasing use of firearms in the seal hunt had led to further decline of the seal populations, particularly because the hunts had shifted towards seals in the sea and not on land. These pelagic seal hunts constituted the mainstay of the hunts carried out by the ACC and amounted to approximately 25,000 animals per year. Towards the end of the 20-year lease, the ACC no longer obtained a renewal. Instead, the lease was given to the *North American Commercial Company*, which hunted around 17,300 seals per year, thus significantly less than the ACC. In 1910, however, the leasing system was abandoned.

Despite this exclusively domestic regulation, a clear international development in regard to sealing took place in the Bering Sea. Throughout the 1880s, the US increased its enforcement in Alaskan waters. After all, pelagic sealing was continuously on the increase and throughout the 1890s would increase to a staggering 42,000 seals per year. These hunts, however, were not conducted by licensed companies, but carried out by vessels from Great Britain, Russia, Japan, Germany, Norway, France and Sweden. In 1887, armed customs vessels seized an American and a Canadian (British) ship for illegally hunting seals in Alaskan waters. After long negotiations, both the governments of the United States and of Great Britain agreed in 1891 to prohibit vessels sailing under their respective flags to hunt seals in the Bering Sea altogether for the period of 1 year. This is commonly referred to as the *modus vivendi* in the Bering Sea seal hunts, given that no official document or treaty exists that clearly lays out this prohibition.

Despite the existence of the *modus vivendi*, the issue of the ships seized by US Customs had still not been resolved. Although the negotiations for an extension of the *modus vivendi* were ongoing, an arbitration procedure concerning the seized vessels before the Court of Arbitration, which was launched in 1892 and concluded 1 year later in Paris, superseded any subsequent *modus vivendi*. The Award Between The United States and the United Kingdom Relating to the Rights of Jurisdiction of United States in the Bering's Sea and the Preservation Of Fur Seals (Arbitration Treaty) was concluded on 15 August 1893.[1] This treaty established a sanctuary of 60 nautical miles (approximately 111 km) around the Pribilof Islands and therefore constituted a direct challenge to US sovereignty within its own territory. Contrary to the Jan Mayen Treaty, this bilateral international agreement had a direct effect on the way the US was entitled to regulate sealing activities in US waters. Following previous regulatory practice, the Arbitration Treaty furthermore established that despite these provisions the native population was still allowed to hunt seals provided their products were not used for commercial purposes.

Since the Arbitration Treaty was a matter between the United States and Great Britain, other countries, such as Russia or Japan, which were also engaging in sealing in the Bering Sea, but that were not part of the process of arbitration, were not obligated to follow it. While, therefore, US and British pelagic sealing takes went down, other countries intensified their efforts significantly, prompting the United States to convene an International Fur Seal Conference in 1897, which was attended by the US, Russia and Japan. There, the Convention for the Preservation of the Fur

---

[1] Award Between The United States and the United Kingdom Relating to the Rights of Jurisdiction of United States in the Bering's Sea and the Preservation Of Fur Seals, 15 August 1893.

Seal and Sea Otter in the North Pacific Ocean and Bering Sea[2] was adopted. Since a precondition for its entering into force was Great Britain's ratification, which never chose to do so, it remained a document that was merely voluntary. Irrespectively, the three other countries chose to adhere to the convention for the 1898 sealing season. The objective of the convention—the cessation of all sealing and sea otter hunting operations in the North Pacific Ocean, including the Bering, Okhotsk and Kamchatka Seas for 1 year—was implemented by Russia, the US and Japan. Since Great Britain (and Canada) refused to follow suit, towards the end of the nineteenth century, the Bering Sea was still the ground for extensive pelagic seal hunts.

6.3 The First Half of the Twentieth Century

The early twentieth century was marked by large-scale seal hunts in the North Atlantic and in the North Pacific. Additionally, regions such as the White Sea, the North and Baltic Seas as well as in Lake Ladoga, the freshwater lake that the then-Russian/Finnish border cut in half, were important hunting grounds. While in the North Atlantic, North Pacific and White Sea the hunts were conducted primarily for commercial purposes, the Baltic Sea and Lake Ladoga hunts were rather a means of controlling seal populations. The overall approach to seals and sealing was therefore significantly different: while in the Bering Sea seal hunts were multilaterally regulated in order to prevent an overexploitation of seals, in European waters each country regulated the hunt itself, even putting in place bounties for seals. In other words, in the Baltic or North Seas, no international agreement existed that led to the regulation of sealing or the protection of seals. Lake Ladoga is a somewhat different story, however, albeit for reasons that are related to Russian-Finnish relations and the fact that Lake Ladoga was a border lake. The same can be said about the Russian-Finnish maritime border area in the Arctic Ocean. Before the Second World War, Finland's territory extended to the northernmost reaches of the Scandinavian Peninsula—the Petsamo Region—which is now part of Russia. Given the ongoing seal hunts in that region as well, Russia and Finland entered into an agreement, yet, once again, not based on sealing *per se*, but on the fact that this was a border region. We will get to this issue below.

6.3.1 The 1911 Fur Seal Convention

Let us first turn back to the Bering Sea, which was subject to an emerging regime on the seal hunt between several sealing nations. After the failure of the International Fur Seal Conference and the not entering into force of the Convention for the

[2]Convention for the Preservation of the Fur Seal and Sea Otter in the North Pacific Ocean and Bering Sea, 6 Nov 1897.

Preservation of the Fur Seal and Sea Otter, it was the United States which took stronger unilateral steps to limit sealing in Alaska. In order to make the hunt for females unprofitable, the US marked fur from female seals in order to devalue their pelts. In addition, US citizens and companies were prohibited from hunting in the open sea north of 35° north latitude. Other countries, however, did not follow this example and the pelagic seal hunt continued to thrive in the Bering Sea, much to the detriment of the fur seal populations, which came closer to collapsing than ever before. In order to put a halt to this development, in 1905 the United States took the initiative and approached the governments of Russia, Japan and Canada/Great Britain again in order to develop a regime for the protection and utilisation of Northern fur seals.

Also Russian officials saw the need for such agreement and by 1908 Russia, Japan and the United States agreed that there is a need for a legal regime protecting Northern fur seals. Great Britain, however, was inert in this regard while also Canada—despite not being an independent country yet—also expressed that it would enter into such an agreement. A few years still passed, but behind the scenes unofficial negotiations had already begun. This resulted in a bilateral agreement between Canada and the United States in February 1911 in which both countries agreed to stop pelagic sealing once also the other two countries would agree to do so. Three months later, the United States hosted a second conference on Bering Sea sealing, this time, however, with the aim to abandon pelagic sealing altogether. A major obstacle that stalled the reaching of an agreement was the issue of compensation—an issue that had already arisen during the first Fur Seal Conference in 1897. While that year it was Canada that raised this concern, in 1911 it was Japan which noted that it wanted to be adequately compensated for not being able to conduct and benefit from sealing. Disagreements abounded to a degree that prompted US President William H. Taft to contact Japanese Emperor Mutsuhito (Meiji) directly on 12 June 1911—a highly unusual step—pleading for an agreement if also the US delegation was to make some concessions. Five days later, Emperor Meiji responded, agreeing to instruct also his delegation to make concessions. As a result, the conference successfully concluded on 7 July 1911 with the adoption of the Convention Respecting Measures for the Preservation and Protection of the Fur Seals in the North Pacific Ocean (Fur Seal Convention).[3] Contrary to its 1897 predecessor, it did enter into force on 15 December 1911.

The main objective of the Fur Seal Convention, as its full title reveals, is the adoption of "effective means for the preservation and protection of the fur seals."[4] To reach this objective, all pelagic sealing, which had ultimately been recognised as being the main cause for the continuous decline of the fur seal populations, was to be prohibited in the North Pacific for the citizens of the parties as long as the convention remained in force. The only exception to this full prohibition was the native

[3]Convention Respecting Measures for the Preservation and Protection of the Fur Seals in the North Pacific Ocean (Fur Seal Convention), 7 July 1911.

[4]Fur Seal Convention, Preamble.

population, consisting of native Americans, Aleuts, Ainu and other indigenous peoples who lived in the convention area. The precondition for their continuing pelagic sealing was a non-commercial nature of that hunt. In addition to a hunting prohibition, the Fur Seal Convention furthermore established trade barriers for skins of the Northern fur seal stemming from the populations of that could be found in the Americas, Russia and Japan.

Given that compensation for the economic losses stemming from the abandonment of sealing played such a defining role during the negotiation process of the convention, a rather detailed outline of how and to what percentage compensation is to be paid could be found in the convention. This effectively meant that even though the Fur Seal Convention had entered into force, this did not mean that no seals were killed anymore. Instead, the seal hunts had moved to the land areas in the Bering Sea, yet strictly controlled by the respective governments. Of the respective revenue stemming from the hunts, both the US and Russia surrendered 15% to Japan while Japan, in turn, surrendered 10% of its herd to the other three governments. Canada also received 15% compensation from Russia. At the time of the convention's conclusion, however, Canada had already ceased its sealing operations in the Bering Sea and a compensation was therefore somewhat unjustified. Also the other parties merely grudgingly complied to this provision in order to maintain friendly relations and to ensure the longevity of the convention. Indeed, the compensation mechanisms are of such fundamental importance that the efficacy of the convention was inevitably tied to them.[5]

Even during the Great Depression of the 1920s the Fur Seal Convention endured. During the time of the convention's being in force, the seal populations in the Bering Sea recovered, which proved to be a direct result of the effectiveness of the convention. The parties, it appeared, had managed to establish a regime that appeased all players involved and benefitted the seal. Further reaching geopolitical developments took their toll on the convention, however. The outbreak of the Second World War and ultimately the outbreak of the Pacific war between Japan and the US and its allies also led to the collapse of the convention: on 23 October 1941 Japan officially withdrew, which meant that the regime was no longer in effect. Northern fur seals, it appeared, could consequently be hunted in the sea again. The remoteness of the Pribilof Islands, along with the war efforts of all parties elsewhere in the world led to a notable scarcity of hunters in the early years of the war. In other words, hardly any fur seal hunts were taking place during that time, which even benefitted the fur seal populations even more. But beyond that, the US and Canada—since 1931 an independent country—bilaterally agreed on the upholding of the provisions of the convention even though it no longer was in force. However, the Aleutian Islands were not spared from being battle grounds in the war and after the Japanese occupation, which lasted from June 1942 to August 1943, sealing was resumed by different parties. Despite the recovery of the herds, increasing pressure

[5]Bailey (1935), p. 4.

was put on them again, making the need for a successor of the Fur Seal Convention ever more necessary. We will turn to this successor below.

6.3.2 The 1922 Regimes Between Finland and the Soviet Union

The year 1922 was an important year for bilateral relations between Finland and Russia (the Soviet Union).[6] We must remember that Finland only gained independence in 1917 while in 1922 the Soviet Union was established. A key treaty in this regard was the Treaty of Tartu[7] of 1920, the post World War I peace treaty between Finland and Russia, manifesting the borders between of both countries, which would last until the Second World War, and which would tighten bilateral cooperation. On the one hand, the borders of Finland and Russia of that time differ from the contemporary borders in that sense as the Arctic region of Petsamo—now part of Russia—was part of Finland, which therefore had access to the Arctic Ocean. On the other hand, Lake Ladoga in the South saw the border between both countries, but now belongs fully to Russia. After all, the borders of Finland at that time corresponded to those of the Grand Duchy of Finland within the Russian Empire.

The Treaty of Tartu established that both countries were to give unrestricted access to their respective waterways. This means that also the waters of the Petsamo region, the Barents Sea (Arctic Ocean), as well as Lake Ladoga saw Russian and Finnish fishers actively fishing in the other country's hegemonic waters. The seal hunts were equally subject to this agreement. In the Barents Sea, seal hunts had been going on for several hundred years, particularly as part of the so-called Pomor Trade, a trade that had been established in the seventeenth century along the northernmost coasts of Norway, Finland and north-western Russia. The indigenous Sámi were main providers for marine resources within the trade system. But also within the White Sea, Norwegian and Russian hunts for seal provided important produce for the Pomor Trade, which followed local and regional needs, making it resilient towards political and economic turmoil. However, upon the turn of the nineteenth to twentieth centuries, the Pomor Trade had already declined significantly and was almost non-existent anymore. Nevertheless, the spirit of cooperation also pervaded the Finnish-Russian agreements that would regulate fishing and sealing.

The first 1922 agreement was the Agreement between Finland and Russia on Fishing and Sealing Operations in the Territorial Waters of both Countries in the Northern Arctic Ocean.[8] The agreement followed the spirit of the Treaty of Tartu and enabled access to fishing and sealing grounds for both Finnish and Russian citizens.

[6]Sellheim (2016a).

[7]Peace Treaty between the Republic of Finland and the Russian Socialist Federal Soviet Republic, 14 October 1920.

[8]Suomen ja Venäjän välinen sopimus kalastuksen ja hylkeenpyynnin harjoittamisesta molempien maiden aluevesillä Pohjoisella Jäämerellä [Agreement between Finland and Russia on fishing and sealing operations in the waters of both countries in the Northern Arctic Ocean], 21 October 1922.

Here we must note that reference is made to *Russia* and not the Soviet Union. After all, this agreement was concluded in October 1922 while the Soviet Union was established in December 1922. Yet leaving this aside, the agreement was significantly different to the regime that was established in the Bering Sea. For instance, it did not consider sealing under environmental pretexts, but rather as an administrative activity that aims to deepen cooperation between Finland and Russia. In other words, the agreement made no reference to catch limits or anything else that might have classified sealing as a potential environmental problem. Instead, the agreement focused on technical issues such as proper documentation fishers and sealers have to carry on them when engaging in fishing and sealing in the other party's waters; how their boats were to be marked; or documentation relating to the logistics of fish and seal processing on land. What we have here is therefore not a sealing agreement in the more classical sense—meaning an agreement concluded to be able to continue sealing—but rather an agreement that must be considered in larger geopolitical contexts and a larger legal discourses. Or to put it differently, the focus of this agreement did not rest on the seals or on the sealers, but rather on the deepening of Finnish-Russian cooperation.

The same can be said about the second agreement that was concluded that year, the Finnish and Russian Agreement on Fishing and Sealing Operations in Lake Ladoga.[9] Lake Ladoga is a freshwater lake in which the Ladoga ringed seal (*Pusa hispida ladogensis*) can be found. The Ladoga ringed seal is one of only four freshwater seal species that exist in the world: the Saimaa ringed seal (*Pusa hispida saimensis*) in Finland, the Baikal seal (*Pusa sibirica*) in Russia, and the Ungava harbour seal (*Phoca vitulina mellonae*) in Canada. Despite their special adaptation to fresh water, local fishers never paid much attention to such genetic forte. Instead, seals constituted a direct competitor and contributed to the destruction of fishing nets and other damage to fisheries operations. Therefore, as was common practice in the Baltic Sea, the Finnish government paid bounties to fishers in their pursuit of seals in order to protect the fisheries. This practice was also established by the Soviet government in the 1920s and the main goal of this effort was the elimination of the Ladoga ringed seal population. After all, its main forage were salmon and whitefish, which were also the main species fished by Soviet fishers. Consequently, the seal was named an 'Enemy of the fisheries' and the Soviet propaganda machinery did not spare it from depicting it as a 'harmful beast' that needed to be exterminated.

With this in mind it is not surprising that the Ladoga agreement, as its Arctic Ocean cousin, did not hold an environmental dimension, but was also fundamentally technical, essentially establishing fishing sealing rights for both parties in each other's waters. The main difference to the Arctic Ocean agreement was that it merely held a 10-year life span. Why this was inserted cannot be ascertained. The establishment of fishing and sealing seasons, lasting from 1 May to 15 November, could be

[9]Suomen ja Venäjän sopimus kalastuksen ja hylkeenpyynnin harjoittamisesta Laatokalla [Finnish and Russian Agreement on Fishing and Sealing Operations in Lake Ladoga], 28 October 1922.

interpreted as an environmental provision, but it remains doubtful whether it referred to the whelping seals or to hatching fish.

Leaving this aside, a slight shift towards environmental concerns occurred in the successor of the agreement, which was concluded in 1934. Even though it was merely marginally different to its predecessor, it nevertheless contained a prohibition on indiscriminate hunting techniques. This meant that it was no longer permitted to apply explosive, toxic or narcotic methods. The 1922 agreement, on the other hand, merely referred to "authorised hunting methods" yet without further definition. Be that as it may, both agreements lasted until the outbreak of the Second World War and became obsolete with Finland's loss of its Arctic Petsamo region and the eastern parts of Karelia, established in the 1944 Moscow Armistice and ultimately enshrined in the 1948 peace treaty between Finland and the Soviet Union.

6.4 The Second Half of the Twentieth Century

By the end of the Second World War and in the years of geopolitical and legal reconstruction of the global order, the groundwork for future sealing agreements was laid. This is to say that sealing was an important part of the international economic system, yet with primary importance for local and regional populations. This is best exemplified by the large-scale commercial seal hunts in Newfoundland, Canada, which prior and after the Second World War provided important economic opportunities for the British colony (which would become part of Canada in 1946).[10] As mentioned before, the main difference to the whaling industry was that sealing was not based on *global* commons and thus that international sealing efforts were carried out by a limited number of nations focusing on specific regions and waters with the associated implications of sovereignty.

Before the Second World War and shortly after, seals were nevertheless globally perceived as an important contributor to the economies of specific regions. Since the discovery of seal oil as a cheap alternative to whale oil, seal oil played an important international role, particularly for the European markets. With the advent and quick advancement of fossil fuels, however, this importance diminished significantly and focus shifted towards seal fur as a fashion commodity. Both seal oil as a fuel source and seal fur for fashion markets contributed to significant declines in seal populations in the Northern Hemisphere and was thus a primary motivator for the emergence of the Bering Sea fur seal regime. After WWII, also other countries started to follow suit. Particularly Norway stands out in this regard, entering into two bilateral agreements with the Soviet Union and Canada respectively, as we will see below.

[10]Busch (1985).

6.4.1 The 1957 Interim Convention on Conservation of North Pacific Fur Seals

Understandably, directly after the war, countries were not overly concerned with the consideration for the environment and ultimately for the conservation of seal stocks. This did not necessarily account for the United States and Canada with regard to the Bering Sea fur seals, however. As we have seen above, despite the ongoing war efforts and post-war restoration initiatives, the two countries pledged to uphold the provisions of the 1911 Fur Seal Convention in order to protect the seal herds. With the advent of the 1950s and the setting of the international world order, the United States did not only take the lead in re-shaping and establishing international institutions, but also in becoming a forerunner in the protection of marine mammals. Since the US had already previously demonstrated that it was possible to unite sealing parties in the Bering Sea under one regulatory regime, by 1951 the State Department had started its work on the process of renegotiating the Fur Seal Convention with Japan and the Soviet Union. When the US, Japan and many others signed the Treaty of San Francisco, the peace treaty with Japan, in September 1951, the path for Japan's involvement in a new Bering Sea fur seal regime was paved.

The first step into the direction of a new regime was taken when the US, Canada and Japan entered into a non-binding cooperation agreement, somewhat similar to a Memorandum of Understanding, to carry out joint research on population statuses, migratory behaviour and feeding habits of fur seals in the Pribilof Islands. The overall intention was that this joint research programme would ease the conclusion of a new fur seal convention. Although also the Soviet Union was approached to join this programme, it refused. While this was the case, the Soviet Union indicated that it would be willing to join a new fur seal convention. Why the Soviet Union did not wish to enter into the agreement on research cannot be explained, nor its wish to directly move towards a binding convention. What is obvious, however, is the fact that by the early 1950s the demand for fur from the Bering Sea began to dwindle. In other words, the commercial aspect of seal hunting in the Bering Sea had lost importance and indeed by 1954 the Northern fur seal was commercially no longer attractive.

This fact may have made it easier for the parties concerned to enter into a regulatory regime for fur seals. For 1 1/2 years, starting in November 1955, the four states met regularly in Washington DC, to discuss and draft a new fur seal convention. Finally, a text was agreed upon and the Interim Convention on Conservation of North Pacific Fur Seals[11] was adopted on 9 February 1957. It entered into force in October of the same year. It would expire on 14 October 1984. The Interim Convention was a treaty that linked several aspects that are still relevant today and that still fuel the controversy surrounding marine mammal exploitation: conservation, sustainable use and research (science). This threefold objective could be found in the Preamble of the convention which aimed to achieve maximum sustainable

[11] Interim Convention on Conservation of North Pacific Fur Seals, 9 February 1957.

productivity of the 'seal resource' while at the same time maintaining a sustainable conservation status and recognising the role of fur seals in the wider ecosystem. In order to achieve this, scientific inquiry and international cooperation would be crucial.

Although seals were considered a resource, their exploitation was consequently not without limits. As in the 1911 convention, pelagic sealing, the seal hunt on the open sea, was strictly prohibited. Two exceptions in this regard were possible: first, if pelagic sealing was carried out for research purposes, not exceeding specific numbers for the eastern and western North Pacific Ocean; second, if the Fur Seal Commission permitted pelagic sealing provided that, in combination with land sealing, it would not jeopardise the objectives of the convention. The Fur Seal Commission was an international commission consisting of members of the convention's parties, which would make decisions on research concerning fur seals, the translation of this research into management decisions and whether or not pelagic sealing would be possible without undermining the objective of the convention. While research and conservation stood at the fore of the role of the commission, the underlying discourse was, however, that of the fur seal as an exploitable resource. This was further underlined by the fact that the commission's work was to take into consideration the subsistence needs of the local native population.

Even though the tripartite objective—conservation, sustainable use and research—was significantly different to the 1911 Fur Seal Convention, some elements were directly, if not verbatim, taken over. For instance concerning the aboriginal exemption, this was almost the exact wording as in 1911:

> The provisions of this Convention shall not apply to Indians, Ainos [sic], Aleuts, or Eskimos dwelling on the coast of the waters mentioned in Article III, who carry on pelagic sealing in canoes not transported by or used in connection with other vessels, and propelled entirely by oars, paddles, or sails, and manned by not more than five persons each, in the way hitherto practiced and without the use of firearms; provided that such hunters are not in the employment of other persons or under contract to deliver the skins to any person.[12]

Also the compensation mechanisms, which were indeed a controversial element of the 1911 convention, were taken over. It was thus that the Soviet Union and the United States were to yield 15% of the commercial value of their hunts to Japan and Canada. Why these provisions were inserted cannot be ascertained. Be that as it may, the Interim Convention was not open ended but had an expiry date: once it had been adopted, it was to be in force for 22 years unless a revised convention were to be adopted by the parties.

While throughout the 1960s the Interim Convention remained the primary regime to control the hunt for fur seals in Alaska, the year 1972 marked a legal and normative shift in the United States. We must remember that the environmental movement took off throughout the 1960s and that the exploitation of marine

[12]Interim Convention, article VII; See also article 4 of the 1911 Fur Seal Convention.

mammals, particularly whales and seals, faced significant resistance in the western hemisphere. Calls for a whaling moratorium at the 1972 Stockholm conference as well as the growing on-the-ground protests against sealing in Newfoundland are but the tips of the iceberg of a changing international discourse on marine mammals and marine mammal hunts. As we have seen above, the United States has been a forerunner in efforts to conserve fur seals and has been the driving force in establishing the fur seal regime. While international efforts to protect fur seals were crucial in the first and early second half of the twentieth century, the domestic legal sphere would change dramatically in 1972 when the US adopted its Marine Mammal Protection Act (MMPA). The act establishes a blanket prohibition on the killing, taking, import and export of marine mammals. The only exception in this regard are native peoples, provided they do not engage in the hunts for commercial purposes. This explains the United States' strong support for its native population, i.e. the Inuit (Eskimo) of Alaska and the Makah of Washington State, at the IWC.

The adoption of the MMPA was the manifestation of a changed discourse in the United States. While the Interim Convention was rooted in the notion of conservation *and* sustainable use, the MMPA merely focuses on conservation irrespective of any possibility of sustainable exploitation. Consequently, the concept of maximum sustainable yield (MSY) was replaced by an ecosystem-based management approach. Despite the fact that negotiations were started again in 1979 for a successor to the Interim Convention, the practice of the United States of inserting a no-use approach to all marine mammals into its own legal environment violated the spirit of consultation and cooperation on the fur seal issue. All research programmes that aimed towards a potential reopening of the commercial seal hunt in the Pribilofs were effectively terminated when in 1983 Congress passed the Fur Seal Act Amendments.[13] By then, therefore, the Interim Convention had lost its momentum and officially became void in 1984.

6.4.2 Norway, A Key Player of for Seal Conservation and Utilisation

The Bering Sea's Interim Convention was not the only agreement that was concluded in the year 1957. Also the second region of international seal hunting efforts, the Northeast Atlantic, saw the conclusion of an agreement between its main actors: Norway and the Soviet Union. Both countries conducted rather intense seal hunts both at Jan Mayen Island as well as in the Barents Sea. As we have seen above, these regions were subject to previous regimes and it is therefore not surprising that, given their importance as seal hunting grounds, further international, in this case bilateral, efforts were necessary to protect and use the prevailing seal populations. Consequently, in November 1957 Norway and the Soviet Union adopted the

[13]An Act to provide for the orderly termination of Federal management of the Pribilof Islands, Alaska. (Public Law 98-129), 14 October 1983.

Agreement on Measures for Regulating the Catch and Conserving Stocks of Seals in the Northeastern Part of the Atlantic Ocean.[14]

Fundamentally, the agreement mirrored the approach of the Bering Sea Interim Convention: on the one hand it strongly fostered the sustainable use of seals; on the other it focused on a scientific approach concerning management advice. This means that the overall objective of the agreement was maintaining of seal populations to ensure a maximum sustainable catch as well as the extension of scientific research in order to determine the condition of seal stocks in the agreement area. However, while the Bering Sea regime focused merely on Northern fur seals, the Norwegian-Soviet sealing agreement took into account three seal species: the Greenland (harp) seal (*Phoca groenlandica*/*Pagophilus groenlandicus*), the hooded seal (*Cystophora cristata*) and the walrus (*Odobaenus rosmarus*). The agreement was designed in a way to be able to also include the bearded seal (*Erignathus barbatus*) and the ringed seal (*Phoca hispida*). Most interestingly, however, the agreement was potentially also expandable to include the polar bear. Whether this has ever occurred remains in the realm of speculation. The agreement is therefore not a regime that is solely concerned with seals, but takes a wider approach to ice-dependent marine mammals.

The agreement area was clearly defined in the Annex, spanning from the Greenland Sea and Denmark Strait to the Barents Sea. In order to set specific hunt limits, a commission was established, which consisted of three member of each party and which was responsible for providing management advice to the parties, to submit proposals for further research and to oversee the implementation of regulations concerning seal hunts.

The commission was therefore the 'heart and soul' of the agreement and the crucial element in its implementation. The decision-making power rested nevertheless still with the nation states. After all, the commission provided merely advice to governments and did not hold decision-making power on its own. The national government was therefore able not to follow the advice of the commission and to pursue its own agenda. In how far this would undermine the entire *raison d'être* of the agreement is, of course, a matter of speculation. Nevertheless, the national governments were therefore also the entities that the agreement entitled to issue sealing licences irrespective of the management regulations that were taken based on the commission's advice.

As in the Bering Sea Convention, these licences applied to seal hunts for scientific research as well as seal hunts conducted by the local (indigenous) population. The parentheses around the term indigenous are quite significant. The licensing provisions were outlined in Article IX of the agreement. In the authentic texts—in Norwegian and Russian—the term 'indigenous' does not occur, but instead the Norwegian term 'stedlige' and the Russian term 'коренно' were used to describe the population in question. The former is rather translated into 'resident' and the latter into 'native' or 'aboriginal'. In the French translation the term 'autochtone

[14]Agreement on Measures for Regulating the Catch and Conserving Stocks of Seals in the Northeastern Part of the Atlantic Ocean, 22 November 1957 (1958 UNTS 280).

locale' was used while the English translation made use of the term 'local indigenous'. While it seems clear which population is made reference to, we do see here differences in the terminology which are quite significant. In the Norwegian text, the denominator of indigeneity did not occur while in the Russian version it did. Whether this had any direct impact on the applicability of the regime cannot be said at this stage and also the literature, which is sparse to begin with, does not make reference to this issue at all.

Be that as it may, despite the fact that the agreement covered areas that did not fall under the direct jurisdiction of a nation state, times were changing in the 1970s and countries started to apply the 200 nautical mile exclusive economic zones, which was manifested in the 1982 UNCLOS. Although the agreement did not have an expiry date, it became void when the UNCLOS was adopted since Norway and the Soviet Union, along with other states, established the EEZ as customary international law. The control over the marine resources occurring within that limit thus rests with the nation states and the need for international cooperation became obsolete.

Although the tides were turning in the 1970s, Norway remained firmly committed to the conservation and utilisation of seals, based on scientific advice. In 1971, it therefore entered into a second agreement, this time with the long-time sealing country Canada. Canada developed an international reputation as ruthlessly decimating harp and hooded seal populations in the North Atlantic, particularly because of its large-scale, commercial seal hunts in Newfoundland and Quebec. Especially throughout the 1960s and 1970s international resistance had developed against these hunts, which were considered to be detrimental to the seal populations and inherently cruel. Particularly the latter narrative would find its way into policy-making in the European Union, which we will get to know later on in the book. It was especially Canada's seal hunts, which contributed greatly to the emergence of the anti-sealing movement and the associated establishment of a public discourse that no longer accepts seals as an exploitable resource. Organisations such as the *International Fund for Animal Welfare* (IFAW) emerged out of resistance against Canadian sealing.[15] Also *Greenpeace* and other organisations made use of the harp seal, and particularly its 'whitecoat' pups, to raise money, despite the fact that the harp seal has never been on the IUCN's endangered species list. In an interview from 1978, *Greenpeace* and *Sea Shepherd* founder Paul Watson underlined that "the seal is very easy to exploit as an image [. . .] which portray(s) a baby seal with the tears coming out of its eyes."[16] Without needing to delve more deeply into this issue at this stage—we will learn more about the roles of NGO in Chap. 7—suffice it to say that Canada was and still is a primary actor in large-scale seal hunts, resisting international pressure and counteracting a strong discourse against sealing.

Despite changing discourse, in July 1971 Norway and Canada entered into the Agreement between the Government of Canada and the Government of Norway on

[15]Barry (2005).

[16]Frum (1978).

Sealing and the Conservation of the Seal Stocks in the Northwest Atlantic.[17] We must remember that Norway and Canada have not had the easiest relationship as regards to sealing. Canadians, and particularly Newfoundlanders, in the past often complained about the Norwegian presence in Canadian waters, operating large vessels and essentially dominating sealing activities. A key provision in the 1971 agreement is therefore that Norwegian sealing is not allowed in the Gulf of St Lawrence and subject to strict geographical limitations in other regions in Atlantic Canada.

First and foremost, however, the agreement had a threefold objective: the conservation of seal stocks; coordination of scientific studies; and the development of humane sealing methods. Particularly the latter marks a significant difference to Norway's 1957 agreement with the Soviet Union and can be clearly ascribed to the changing discourse on sealing, which placed great emphasis on the (in)humaneness of commercial sealing. While the main seal species under the agreement's purview was the harp seal, it was also possible to extend it to include hooded and bearded seals as well as the walrus. To oversee the implementation of the agreement, a commission was established, providing the governments with advice concerning catch limits and other managerial issues as well as concerning humane hunting methods. Compared with the 1957 agreement, the weight of the commission is slightly heavier: on the one hand, the commission was also entitled to submit proposals to the governments concerning inspection and other control measures, enabling a direct influence on enforcement activities. On the other, the commission was to be informed when the national governments issue permits for scientific research, the local population or for expeditions that involve the hunting of seals. While the commission did not have a veto right, the obligation to inform it notwithstanding indicates that its role and opinion are taken into account when permits are issued.

Despite the fact that the agreement was concluded in 1971, it *de facto* did not last very long. After all, in 1977 Canada established its EEZ, which made the work of the commission obsolete. All sealing activities within the 200 nm limit became subject to Canadian regulation. At the same time, Norwegian sealing shifted gradually to Jan Mayen Island while also the increasing role of the International Commission for the Northwest Atlantic Fisheries (ICNAF) in the setting of catch limits for seals in waters beyond national jurisdiction further diminished the commission's importance until it finally ceased to exist.

6.4.3 Antarctic Seals and Sealing

Governance of the Antarctic is taken care of by the Antarctic Treaty System (ATS), consisting of the Antarctic Treaty from 1959, the Convention for the Conservation of

[17]Agreement between the Government of Canada and the Government of Norway on Sealing and the Conservation of the Seal Stocks in the Northwest Atlantic, 15 July 1971 (870 UNTS 85).

Antarctic Seals (CCAS) from 1972, the Convention on the Conservation of Antarctic Marine Living Resources (CAMLR)[18] from 1980 and the Madrid Protocol from 1991. As we have seen in Sect. 4.3.3, marine mammals do play a role in the Antarctic governance, however, not directly under the purview of Antarctic institutions. After all, when the Antarctic Treaty was adopted, the whaling convention had already been in force and the Antarctic Treaty parties did not see a point in agreeing upon another separate regime.

The case was different in regard to Antarctic seals. Since whaling and sealing were primary drivers for Antarctic exploration in the first place and up until the 1960s active seal hunts were ongoing in Antarctic regions, the Antarctic Treaty parties considered it an asset to have an agreement in place that would regulate Antarctic sealing activities. To this end, the Antarctic Treaty Consultative Parties (ATCP), those countries that have since the coming into force of the Antarctic Treaty acceded to the treaty but are not amongst the twelve original parties, decided on voluntary interim guidelines for the protection of Antarctic seals in 1964. Four years later, the Antarctic Treaty Consultative Meeting (ATCM) negotiated the first draft for a convention that was presented to the national governments. At the Sixth ATCM in 1970 it was discussed whether a sealing agreement should fall under the Antarctic Treaty or whether it should be considered as a stand-alone agreement. The meeting opted for the second option since this would also allow non ATCPs and non-Antarctic Treaty parties to enter the agreement irrespective of their ongoing research activities in Antarctica—a crucial element in the status of ATCPs. In 1972, therefore, a high-level meeting was held that discussed the draft of the convention from 1970. Ultimately, the meeting adopted the Convention for the Conservation of Antarctic Seals (CCAS),[19] which would put a regulatory regime for Weddell seals (*Leptonychotes weddellii*), crabeater seal (*Lobodon carcinophaga*), Ross seal (*Ommatophoca rossii*), Southern elephant seal (*Mirounga leonina*), leopard seal (*Hydrurga leptonyx*) and Southern fur seal (*Arctocephalus gazella*) in place. The convention entered into force on 11 March 1978.

The CCAS primarily aimed at reducing the seal hunts on land, which had decimated seal populations drastically in the past. Contrary to the North Pacific, pelagic sealing was in principle still permitted. While that was the case, the Annex of the CCAS reads that the parties "agree to take appropriate steps to ensure that their nationals and vessels under their respective flags refrain from killing or capturing seals in the water" except for scientific purposes.[20] However, a complete prohibition of hunts was put in place for the Southern elephant seal, Ross seal and Southern fur seal. For the other three species, total allowable catches (TACs) were put in place while a total ban on any take was inserted from 1 March to 31 August each year. Moreover, six sealing zones were established within the agreement area, which

[18]Convention on the Conservation of Antarctic Marine Living Resources, 20 May 1980 (1329 UNTS 48).

[19]Convention for the Conservation of Antarctic Seals (CCAS), 1 June 1972 (11 ILM 251).

[20]CCAS, Annex 7.b.

would place extra modes of protection for each seal within them in place. This meant that these regions could be closed from sealing for a specific time period of the year, making it illegal to hunt seals despite the overall opening times for sealing. On top of these stringent measures, three Seal Reserves were established under the CCAS, which prohibit any sealing activities around the South Orkney Islands, in the Southwestern Ross Sea and in the waters between Edisto Inlet, Cape Hallett and Helm Point in Eastern Antarctica.

At the time of adoption of the CCAS, Antarctic sealing *de facto* no longer played a role. Therefore, the enabling of sealing under the CCAS was rather a matter of possibility than of responding to actual commercial needs. The same can be said about the possibility to hunt seals for scientific purposes. Only on one occasion in 1986/87 did the Soviet Union make use of this provision, taking several thousand crabeater seals, around 100 Weddell seals, 1 elephant seal and an unknown number of Ross seals. While scientific sealing was permitted, the issue caused significant unrest amongst the parties of the CCAS, particularly since the Soviet Union did not act in a transparent manner. As a consequence, a conference on the efficacy of the CCAS was held in 1988, which saw a tightening of the relationship between the CCAS and the ATS, particularly with regard to the recently adopted CAMLR Convention in 1980. The fear of a reopening of large-scale commercial sealing drove the debates, but ultimately proved to be nothing more than fears only since the Soviet Union signalled that it would not conduct scientific sealing anymore.

The question therefore arises what the ultimate purpose of the CCAS, in light of the non-existing seal hunts in the Antarctic, really is. After all, the convention was concluded without a real existing threat to the species it aims to protect. Two strains of argument can be applied that would explain its existence: on the one hand, the fact that a threat to Antarctic seals did not factually exist at the time of its conclusion makes CCAS ultimately a precautionary regime. Having learnt from the past, it aims to prevent overexploitation before this overexploitation is actually taking place. In this regard, while ultimately being rather conservative by focusing only on seals and not on the Antarctic ecosystem, it is rather progressive since the precautionary approach as an environmental principle was established only at the 1992 Rio Conference. On the other hand, its role can be seen as an overall desire in protecting the pristine Antarctic environment from any threats, making it rather independent from the Antarctic Treaty itself.[21]

Whether or not the CCAS has factually contributed to the protection of seals in the Antarctic is subject to speculation, particularly since the CAMLR Convention and the Madrid Protocol have taken over many aspects of ecosystem-based conservation efforts. Spain's and Pakistan's expression in 2012 that they would wish to accede to the convention can therefore rather be interpreted as a signal towards other ATCPs that they adhere to the ATS in toto rather than a signal of potential sealing activities in the Antarctic. Unfortunately, I was not able to get official statements

[21]Mossop (2013), p. 278.

from the responsible ministries in Spain and Pakistan on this issue. Up to the point of writing, neither Spain nor Pakistan have become party to the CCAS.

The adoption of the CCAS marked the end of an era concerning legislation for sealing: it was the last of its kinds that linked the sustainable use of seals with their conservation—a feature that had marked the sealing regimes of the twentieth century. The resistance against sealing and whaling, paired with the increasing strength of the environmental movement, prompted states to take different routes. This means that the CCAS was the last regime that considered the seal as an exploitable resource. Over the course of the 1970s, the discourse on seals and sealing had changed to such a level that seals had become a species that could not longer be used for human consumption. Paired with concerns over the population status of harp seals—presumably caused by overhunting by commercial sealing in eastern Canada—and the shifting of seals into a human-like stage (anthropomorphisation), public opinion turned against sealing and sealers and pressed states to take action to ban the marketing of seal products. This process, however, would take well into the 2000s before it was an internationally established practice, as we will see in the following section. The first step was nevertheless taken by the European Economic Communities (EEC), out of which would evolve the European Union (EU) in 1993, by banning the trade in products stemming from harp and hooded seal pups in 1983.

6.4.4 Seals in the Mediterranean and in the Wadden Sea

Seals in the Mediterranean and in the Wadden Sea, the southern edges of the North Sea, have been put under different modes of protection. At the core of seal protection in the Mediterranean stands the Barcelona Convention, the 1976 Convention for the Protection of the Mediterranean Sea Against Pollution,[22] which was renamed to be the Convention for the Protection of the Marine Environment and the Coastal Region of the Mediterranean in 1995. In light of the drastic declines of the Mediterranean monk seal (*Monachus monachus*) throughout the 1970s, the parties to the Barcelona Convention, which, as the name implies, primarily dealt with pollution of the Mediterranean Sea, decided to take action. In 1987 the parties initiated an Action Plan for the Management of the Mediterranean Monk Seal which aimed, for instance, to reduce adult mortality and to establish marine protected areas for the seal. However, despite these efforts, one of the primary threats to the species are deliberate and unreported killings by local fishers. Also in 2019, the Mediterranean monk seal remains critically endangered, pointing towards the fact that protective measures have not been successful.

The situation is somewhat different in the Wadden Sea. Here, grey and harbour seals have been hunted since the ninth century. Ultimately, this, paired with other factors, led to significant declines of these species in the area, which led to the

[22]Convention for the Protection of the Mediterranean Sea Against Pollution, 16 Feb 1976 (1102 UNTS 27).

population not even being listed as endemic to the Dutch fauna anymore. While hunting may have had impacts in the past, it was particularly the outbreak of the phocine distemper virus in 1988 that reduced the seal populations drastically. In order to ensure the species' survival, Denmark, Germany and the Netherlands adopted the Agreement on the Conservation of Seals in the Wadden Sea in 1990[23] as one 'Agreement' under the CMS (see Sect. 4.3.2). The agreement fosters the cooperation between the countries on the protection of Wadden Sea seals through the adoption of seal management plans, scientific studies and monitoring. The protective measures in place have led to a significant growth of the populations. Despite this, the deliberate killing of Wadden Sea seals takes place for two reasons: first, fishers consider them pest species since they interfere with fishing activities; second, in order to prevent further outbreaks of diseases, individual seals are being killed to protect the overall population.

6.5 The Shifting Paradigm: The 1983 Seal Pups Directive

The large-scale hunts in Atlantic Canada were to a large degree driven by commercial incentives, meaning that it was the demand for seal products that motivated these hunts in the first place. As we have learnt above, seal oil was no longer of interest to the world since the development of hydrocarbon resources. Instead, it was the fashion industry that was in high demand for seal fur, first and foremost stemming from harp and hooded seal pups. The former, the so-called 'whitecoat', is indeed the perfect poster species that almost everybody will know: a white, fluffy seal with large black eyes. A seal that embodies innocence and cuteness, fragility and purity. On the one hand, this purity spurred the demands from the fashion industry since the white fur was a much sought-after commodity for high quality fashion products; on the other hand, it was exactly these trademarks that fuelled the anti-sealing protests of the 1970s. At that time, it was particularly the European markets that were looking for seal furs, inevitably making them the primary target for anti-sealing protests.

Three countries were particularly interesting for the anti-sealing movement in this regard: the United Kingdom, West Germany and the Netherlands. Organisations such as *IFAW* or *Greenpeace* put significant efforts into creating materials that depicted the cruelty of the seal hunt and that addressed the conservation status of seals. The efforts did not miss their mark and several million letters and phone calls by concerned citizens of Europe reached the European Parliament, putting pressure on policy makers, urging them to take action to protect seals. But letters and phone calls did not constitute the only means of action: in the UK and in West Germany, thousands of people took to the streets protesting against the seal hunt and for the putting into place of trade measures to stop the hunt.

Impressed by such massive public outcry, in 1978 Italy was the first country to nationally ban the import and trade of seal products stemming from seals smaller

[23] Agreement on the Conservation of Seals in the Wadden Sea, 16 Oct 1990.

than 50 cm as per a decree of the Ministry of Commerce. The reasons for Italy's move are somewhat unclear. After all, the country was not a major importer of seal skins, nor was it subject to significant pressure while the Christian Democratic government of Giulio Andreotti did not stand out as being overly concerned with environmental issues. But it set an important precedent for other countries to follow suit and paved the way for a European trade ban on seal products. We must remember that the Italian ban did not address conservation issues nor did it aim to halt the import of seal products stemming from endangered seals. Instead, the ban constituted a blanket ban on *all* products stemming from seals smaller than 50 cm. In two respects the Italian move opened the door for a new era in seal legislation: first, it established the narrative of the seal not being an exploitable resource anymore by putting a blanket ban into place. Second, it followed the notion of species protection through trade measures, which was established by the coming into force of CITES on 1 July 1975.

Although Italy had instigated a first national ban, it was not the first country to take action. Already since 1970 the Netherlands had a voluntary import ban in place, which, however, left it to fur traders themselves to decide whether or not to import seal furs. In how far this voluntary ban was followed cannot be ascertained. But with Italy's nation-wide ban in place, it did not take long for the Netherlands to adopt a similar piece of legislation in September 1980, using conservation concerns as the primary argument. By then, the European discourse on banning the trade in seal products had already reached the level of decision-makers: in April 1980 the first Member of the European Parliament (MEP) from the UK tabled a motion for a resolution calling for a Europe-wide import ban on seal products. While conservation still played a role in the Italian and Dutch bans, the motion referred to the killing of young seals arousing public outrage. Only as a secondary reason conservation concerns were mentioned.[24] Interestingly, however, despite reference to public outrage, the motion mentioned animal welfare aspects of the seal hunt only in passing. The 1970s and early 1980s were the time when it was not conservation, but animal cruelty that drove the discourse on sealing. But while that may have been so, the committee of the European Parliament responsible for the drafting of a resolution on the seal trade referenced concerns over animal welfare and conservation. In other words, European policy-makers identified animal welfare shortcomings and conservation concerns as the prime motivators behind a European legal instrument. In other words, this legal instrument was supposed to improve animal welfare in the seal hunts and it was supposed to have positive effects on the conservation status of seals. Therefore, detailed information on hunting methods and population statuses of seals in Canada and Norway—where also major seal hunts were taking place—were provided and it was concluded that "[i]n view of the threat to the species and the barbaric hunting

[24]European Parliament (1981), p. 5.

methods employed a Community import ban on the skins of both hooded seals and harp seals [. . .] seems appropriate."[25]

Although by early 1982 particularly the Canadian government lobbied strongly for an abandonment of the legislative process to adopt a ban, to demonstrate the humaneness of the Canadian seal hunt and to show that the seal hunt was sustainable, the European wheel of policy-making was already moving. One reason why the Canadian efforts were not fruitful was the fact that Canada raised its TAC for harp and hooded seal for the 1983/84 based on new scientific data. Given that the Europeans were also concerned over the conservation status of seals, this proved to be counterproductive in regard to the negotiations with Europe. While apparently strictly science-based, this did not convince European decision-makers, who were, in fact, in the process of trying to convince Canada and Norway to phase out all sealing activities altogether. Not surprisingly, both countries did not agree. Instead, Canada suggested that the International Council for the Exploration of the Sea (ICES) (see Chap. 4) was to be consulted in matters relating to conservation. Indeed, the European Commission was of the opinion that ICES was a legitimate authority in regard to the population status of seals. However, it would not move away from any concerns concerning the killing methods of seals. The Commission noted that even though seal killing methods were quick and would not cause unnecessary pain to the animal, it was the mere fact that it was seals that were killed which would cause public outrage. In other words, neither conservation nor animal welfare but principal opposition to the killing of seals motivated the legislative process concerning the seal trade.

Throughout the year 1982 Canada tried to argue along science-based lines and thus to avert the adoption of some kind of ban on seal products in the European Communities. These attempts were countered by anti-sealing organisations such as the *IFAW* and *Greenpeace* that themselves used their influence to lobby European institutions. While, on the one hand, they put soft pressure on Brussels by referring to public opinion, they furthermore put hard pressure on national governments, threatening them to publicly denounce them if they were unsupportive of a European seal ban. In October 1982, the European Commission thus presented its first draft for a directive that would ban the import of products stemming from harp and hooded seal pups. Just a little over one month later, ICES published its report on the conservation status of marine mammals which did not support any claim of decline of harp and hooded seal populations due to overhunting. A European ban on products from harp and hooded seal pup based on conservation concerns would therefore not be justified. Irrespective of this finding, the European Parliament approved the Commission's draft proposal and it was now up to the Council of the European Communities to adopt the directive.

In order to avert this adoption, Canada followed its science-based line of argumentation and furthermore requested that the European Commission would start the negotiations on a Canada-initiated, strictly science-based sealing convention, which

---

[25]European Parliament (1981), p. 21.

would include scientific approaches to seal killing methods. The European Commission, however, refused to do so. Not surprisingly, therefore, on 23 March 1983 the Council adopted the first piece of legislation that would specifically remove seals pups from the list of exploitable resources. Although this 'Seal Pups Directive'[26] did indeed consider the exploitation of seals as a 'legitimate' occupation and livelihood and while it was to be in place for merely 2 years as a means for seal populations to recover—even though ICES did not ascribe them worrisome conservation statuses—these provisions appeared to be only half-hearted. On the one hand, the negative effects of the Seal Pups Directive were not reason enough to make changes to it. Both commercial sealers and indigenous subsistence sealers were struck hard by the implications the ban had on the markets. Although the Directive included an exemption for Inuit sealers, Arctic communities were hit by the collapsing markets and Inuit sealers had no longer the opportunity to place their products on the European markets.[27] At the same time, the loss of labour and security for commercial sealers and those working in the processing industry were not considered. On the other hand, these aspects did not prevent the Seal Pups Directive to outlive its set expiry date: in 1985 it was extended for another 4 years until 1989. Then it was extended indefinitely. Legally speaking it is still in force today.

6.6 The 2000s: Banning the Seal Trade

The adoption of the Seal Pups Directive marked an important shift in the way seals and seal hunting were treated legally. The 1980s and well into the 1990s also legally established the seal as a non-exploitable resource, meaning that the lethal taking and utilisation of seals was no longer considered legitimate in international discourse. Since there was, and still is, no international body overseeing the seal hunt, it was up to states themselves to decide whether sealing was still a legitimate occupation or not. For sealing states such as Canada or Norway the position was rather clear: both countries are to this day pursuing sealing, resisting international pressure to end it. The situation in other states such as in Europe, Australia and New Zealand, South America or Southern Africa is different. In Europe, which would become the now 28-state encompassing European Union in 1993, seal hunts are still ongoing, albeit for non-commercial purposes. Also in the 1980s and 1990s seals were hunted, particularly in the Baltic Sea or in Scottish waters, aiming to protect the local fisheries. The Seal Pups Directive did indeed ban the trade in products stemming from harp and hooded seal pups, but it did not ban the hunt for seals *per se*. Sealing for marine management purposes was consequently considered to be a necessity while sealing to fulfil market demands was not. This difference in perception on the justification of the seal hunt can still be found today. The same can be said about

[26]Council Directive 83/129/EEC of 28 March 1983 concerning the importation into Member States of skins of certain seal pups and products derived therefrom (OJ L 91, 9.4.1983, pp. 30–31).

[27]Wenzel (1991).

Australia and New Zealand, South American countries as well as South Africa, all of which would introduce legislation that would make the lethal taking of seals (and other marine mammals) illegal in the course of the 1980s and 1990s. The worldwide discourse on sealing, it seemed, would in the end lead to the legal consensus of making commercial sealing an unwanted and unacceptable activity.

Naturally, the world does not function in unanimous ways. Unimpressed by these developments, Canada, Namibia and Norway continued their seal hunts. Particularly Canada is important to mention in this context since it introduced its Marine Mammal Regulations (MMR)[28] in 1993. While very protective of its marine mammals, Canada did not outlaw the taking of them, particularly not seals. The MMR's provisions on sealing are the culmination of national interests and international influence. On the one hand, the MMR protect sealers from unwanted interference by anti-sealing activists through a strict licencing system and through the establishment of a minimum distance of observers to the ongoing seal hunts. On the other hand, the MMR hold provisions on animal welfare, which had been made somewhat more specific compared to its predecessor, the 1964 Seal Protection Regulations.[29] Since the main international focus still rested on Canada and its commercial East Coast seal hunt, the introduction of the MMR did not bode well. Moreover, since TACs were based on scientific findings and the harp seal population had grown significantly, the Canadian government raised the TAC for harp seals form 185,000 to 250,000 in 1995. Naturally, the international community did not approve and as a consequence, organisations such as *IFAW* or the *Humane Society of the United States/International* (HSUS/HSI) increased their presence on the ice and produced well-produced imagery in order to make Canada end its seal hunts. Canada, however, did not yield.

6.7 The European Union as the Forerunner of the New Paradigm

With this we enter the new millennium and the ever-growing notion of sealing belonging to the past, as being outdated, and as no longer being fit for the twenty-first century. But once again, Canada was not impressed. And in the early years of the 2000s, two important events came together that would manifest a new international paradigm: the banning of the seal trade as a weapon in the fight for seals. The first event was Canada's raising of the TAC of harp seals in 2003 that allowed for the landing of 975,000 seals within a time period of 3 years. This TAC was, once again, rooted in the finding that harp seal populations were thriving. Once again, the international community did not respond favourably and calls for international action against Canada's seal hunt grew stronger. The Seal Pups Directive, it seemed,

[28]Marine Mammal Regulations, SOR/93-56, 4 Feb 1993.

[29]Regulations Respecting the Protection of Seals, Statutory Order Regulations 64/443 (Order in Council P.C. 1964/1663), 29 October 1964.

was not enough to affect the Canadian seal hunt to a degree that would bring it to an end. After all, since also Canada made the hunting of whitecoats illegal in 1987, sealers and the markets for seal products had shifted from whitecoat furs to the silvery furs of slightly older harp seals ('beaters'; see Sect. 2.2.1). The European Union became an important market for beater skins and large-scale processing took place in Greece.

With the raising of the TAC to almost 1 million seals for a time period of 3 years, public opinion, fuelled by anti-sealing organisations called for action. Canada, however, responded by attempting to tighten its seal welfare provisions in the MMR. To achieve this, the Independent Veterinarians' Working Group (IVWG) was formed to assess the animal welfare aspects of the seal hunts in 2005. The findings of the IVWG were to be integrated into the MMR and ultimately the so-called 'three-step-process', the stunning, checking and bleeding of a seal, became mandatory under the MMR in 2009. At the same time, across the Atlantic Ocean, the European Union also dealt with the issue of animal welfare, yet in a different manner. By 2006, the EU's new Action Plan on animal welfare[30] was to be adopted and it was criticised by the European Parliament that, in light of the developments in Canada, it did not address the issue of sealing, which would be considered as 'cruelty products'. As a result, the first parliamentary call for a legal response to Canadian sealing was ushered in the form of a parliamentary declaration in September 2006, requesting a ban on trade in seal products.

6.7.1 The Process Towards a Trade Ban

The Declaration of the European Parliament in essence reflected the overall discourse on sealing which had emerged over the last 30 or so years. Irrespective of the conservation status of seals, which was used as a reason for adopting the Seal Pups Directive 23 years earlier, the call for a total ban on all seal products referred to a number of issues, which can no longer be placed in a science-based context. As reasons for the need for a ban, the European Parliament mentions the young age of seals, in essence therefore applying the 'baby seal' narrative and the associated no-use discourse; the large number of seals that are killed, which, according to the Declaration, reduced the seal population by two thirds in the 1950s and 1960s; the lack of economic necessity to conduct seal hunting in the first place; and the claim of 42% of seals being skinned alive, ultimately referring to animal welfare aspects of sealing. As in the Seal Pups Declaration, Inuit seal products were not supposed to be affected by any trade measure.

In how far the claims brought forth in the Declaration are rooted in scientific or other empirically testable findings is difficult to assess. However, it remains clear that at that point, the conservation status of harp seals was no longer an issue. Several sources, such as ICES or also the findings of the Canadian Department of Fisheries

[30]European Commission (2006).

and Oceans (DFO) did not substantiate the claim of a dwindling seal population. The issue of economic necessity is, once again, not empirically testable and it is doubtful whether the claim stems from independent sources. The Declaration notes that merely 5% of the annual income in sealing communities stem from sealing. This is oversimplified, at best. My own field research in sealing communities in Newfoundland has shown that it very much depends on the year in which the contribution of sealing is measured. Commercial sealers are first and foremost fishers, meaning that their annual income comprises of revenues stemming from fishing and sealing. In simple terms we can say that if the markets for fish are bad and the markets for seal products are good, the annual contribution of sealing becomes much larger. Vice versa, it can, indeed, only amount to 5%. A general claim of ascribing commercial sealing a lack of economic necessity leaves out market fluctuations and, additionally, leaves out other needs, such as cultural and social needs. Lastly, the claim of 42% of seals being skinned while conscious is a gross misinterpretation of veterinary data and appears to be rooted in the desire to justify a EU-wide trade ban based on emotional responses to such claim. Also the European Food Safety Authority (EFSA) concluded in a report on animal welfare aspects of seal hunting that the number brought forth in the Declaration did not correspond to scientific findings.[31]

Irrespective of the problematic claims in the Declaration, which nobody appeared to have 'fact-checked', the Commission initiated the legislative process for the adoption of a trade ban on seal products. At first, the process focused on the animal welfare aspects of sealing and the EFSA was tasked to provide an overview of the findings of veterinary studies and of other materials not collected by itself that would allow for an idea of how (in)humane commercial sealing was. The EFSA concluded that even though in most instances seals are killed swiftly, humane killing does not always occur. Yet it abstained putting a number on this claim. One important recommendation the EFSA put forth was the requirement for the three-step-process—stunning, checking, bleeding—to improve the animal welfare aspects of sealing. This corresponded to the findings of the IVWG and the Canadian MMR followed this request in 2009. Shortly after the report from the EFSA, a Danish consultancy firm, COWI, took over the task of conducting an impact assessment on different policy options. Throughout the report, which was released in 2008,[32] several policy options were tested as regards their environmental, economic and social impacts. As the least detrimental option for the local population was considered a labelling system that would clearly mark those seal products stemming from hunts with high animal welfare standards *vis-à-vis* those for which these standards cannot be proved. This system, COWI concluded, would also improve the image of seal hunting in the world. A total trade ban was considered to be the most detrimental option for Inuit and non-Inuit sealers alike, because buyers would not be able to distinguish between Inuit and non-Inuit products despite the existence of an Inuit exemption.

[31]EFSA (2007), p. 46.

[32]COWI (2008).

When in July 2008 the European Commission presented its first draft for a regulation banning the trade in seal products to the European Parliament, it highlighted the ethical concerns of the European public over the hunting of seals. This finding was based on a questionnaire that was part of the COWI study, but which COWI considered, for various reasons, unrepresentative. Instead of following COWI's advice to implement a labelling system, the Commission proposed an import/export ban as well as a ban on intra-Community trade. From a legal perspective, the Proposal saw the harmonisation of the EU's internal market as the driving force. After all, at that point, the internal market was fractured in regard to seal products since different EU member states had different rules in place.

The different EU institutions received the Proposal with mixed reactions. While some considered it not in line with EU law, others saw it as not strict enough. When the Proposal was finally amended and tabled for a vote in the EU Parliament, its provisions had been tightened. One element is noteworthy here: the Proposal saw a continuation of sealing if high animal welfare practices are adhered to. This provision was removed, indicating that the EU would adopt a measure that did not consider sealing as a legitimate activity anymore. The justification of the Parliamentary committee responsible for the amendments (the Internal Market and Consumer Protection Committee, IMCO) was that commercial sealing is inherently inhumane. In May 2009, Regulation (EC) No 1007/2009 of the European Parliament and of the Council on Trade in Seal Products[33] was adopted with an overwhelming majority. It was followed 1 year later by an implementing regulation. These two regulations are commonly referred to as the 'EU Seal Regime' or, in common parlance, the 'EU Seal Ban'.

The EU Seal Regime places a blanket ban on the trade, import and export of products of *all* seals. This means, any commercialisation of seal products within the market of the EU are prohibited. In its original form, three exceptions to this ban were possible: first, trade was still possible when the products stemmed from indigenous/Inuit subsistence sealing ('Inuit exemption'); second, import was possible when the products were in the personal luggage/property of travellers, indicating non-commercial importation; third, trade was possible when the products stemmed from hunts conducted for marine management purposes, such as seal culls for fisheries protection. This trade, however, was to take place on a non-profit basis ('MRM exception').

In 2015, the scope of the EU Seal Regime was altered, resulting from the proceedings before the World Trade Organization, which I will deal with below. In its amended version, two important changes were made. First of all, the possibility to conduct non-profit trade in marine management seal products was abandoned. Seal products from these hunts can therefore no longer be placed on the market. Second, although the Inuit exemption still persisted, new criteria required high animal welfare standards in these hunts. In how far this provision can effectively

[33]Regulation (EC) No 1007/2009 of the European Parliament and of the Council of 16 September 2009 on Trade in Seal Products, OJ L 286, 31.10.2009, pp. 36–39.

Fig. 6.2 The adoption process of the EU Seal Regime © Nikolas Sellheim

be implemented in light of the remote conduct of Inuit hunts, and in how far this contradicts international standards of indigenous rights, goes beyond the scope of this book.

A schematic of the adoption process of the EU Seal Regime can be found in Fig. 6.2.

6.7.2 Legal Challenges of the Seal Regime

With the Seal Regime in place, it does not come as a surprise that resistance emerged. After all, the European Union was a primary market for seal products and even though Inuit were legally exempted from any trade barrier, they were *de facto* strongly affected by it. This stems from the closely intertwined trade pathways of both commercial seal products and seal products from subsistence hunts. Since no label was in place that clearly distinguished commercial from subsistence seal products, buyers abstained from buying seal products altogether.

Two institutions played a key role in the challenges of the regime: the European Court of Justice (ECJ) and the World Trade Organization (WTO). The former saw several court cases initiated by Inuit sealers and the Inuit Tapiirit Kanatami (ITK), which represents more than 60,000 Inuit, as well as commercial sealing organisations. The ultimately unsuccessful cases to overturn the ban dealt with specific provisions of European law, the complexities of which exceed an introductory book such as this. I therefore spare you from the details. In a nutshell, the Inuit failed to have their claim of being directly concerned by the regime legally recognised. Two detailed analyses of these cases can for example be found in Dorothée Cambou's analysis from 2013[34] and in my own from 2015.[35] I refer to these studies for further information on the legal contexts.

The second proceeding was before the WTO, which was also initiated in 2010. Here, Canada (and later on also Norway) challenged the regime as being in breach of the free trade provisions of the General Agreement on Tariffs and Trade (GATT).

[34]Cambou (2013).

[35]Sellheim (2015).

The EU, however, argued that given the possibility of imposing trade barriers for moral reasons, the regime corresponded to the European moral standards pertaining to sealing. Once again, the case is inherently complex. Two studies that showcase this complexity were conducted by Howse and Langlille in 2012[36] and by Nielsen and Calle in 2013.[37] The result of this case was that the WTO indeed saw the EU Seal Regime in breach of the free trade doctrine since the EU had not demonstrated that the regime in fact corresponds to this moral standard. The argument was that, first, animal welfare considerations did not play a role in the regime—after all it puts in place a blanket ban—and therefore a moral claim could not be made. Second, the seal trade was still possible resulting from marine management hunts, also without any consideration of animal welfare. As a result, the EU was ordered to amend the regime accordingly. Consequently, in 2015 an amended version was adopted in which the MRM exception had been removed while now the Inuit exemption was complemented with high animal welfare standards. Up to the point of writing, no further challenges to the regime have been made.

6.8 Summary and Conclusion

International agreements for the regulation of sealing and management of seals were, up to the 1970s, driven by the incentive to exploit them. All of the regimes that were part of the pages above that were concluded before the 1970s considered seals as a resource that can be used for human benefit. Especially the Bering Sea Fur Seal Regime stands out in this regard, particularly since, in its various forms, is the oldest multilateral cooperation on marine mammals in the world. After all, taking into account the Arbitration Treaty of 1894, it lasted for almost 100 years, pointing towards both the importance of Northern fur seals for the hunting parties involved as well as towards the need of international cooperation in order to adequately manage them. While seals were the focus of this cooperation, they were also part of a larger narrative, particularly between Finland and Russia/the Soviet Union. Seals contributed to strengthen the countries' ties in regard to fisheries and access to each other's national waters. This, in turn, fostered overall cooperation in the spirit of the Treaty of Tartu.

After the Second World War it was particularly Norway that had emerged as a nation entering into agreements on sealing, thereby setting important international standards on scopes and characteristics of sealing laws. Canada and the Soviet Union were important partners in this regard. Notably, these three countries have, principally, maintained their pro-utilisation stance on marine mammals up to the present day. The Russian Federation, however, has tightened its rules on sealing, especially with the imposition of the 2011 trade ban. Within the IWC, however, the Russian Federation still maintains its pro-sustainable use attitude. Whether the imposition of

[36]Howse and Langlille (2012).

[37]Nielsen and Calle (2013).

the seal trade ban marks a larger paradigmatic shift in the Russian position is too early to ascertain. However, the EU Seal Regime may in fact be a direct reason for this shift.

The adoption of the EU Seal Regime has set important precedents for the international trade in seal products. While apparently rooted in the European public's opposition to sealing, it remains questionable in how far it contributes to higher animal welfare standards in the seal hunt. While, naturally, with reduced markets also the number of seals that is being killed decreases, the individual seal that is still subject to seal hunts in Canada or elsewhere does not benefit from it. This is to say that the regime does not motivate commercial sealers to apply higher animal welfare standards other than those provided for in domestic legislation (for example in the Canadian Marine Mammal Regulations). Since this is the case, the EU Seal Regime can hardly be interpreted as a regime based on moral reasons, but instead as a regime that aims to principally oppose, and ultimately end, the seal hunt conducted for purposes other than subsistence sealing. However, as I explained above (see Sect. 3.3), the differentiation between subsistence and commercial sealing is somewhat ambiguous, which points to either lack of knowledge of EU decision-makers on the realities of sealing or to a deliberate ethnicity-based view.[38]

Either way, the EU Seal Regime has not missed its mark. Especially Canadian commercial seal hunts have drastically decreased with the closure of the EU market. Secondly, other countries have followed suit, having implemented similar regimes, taking the EU Seal Regime as a sort of blue print: the Russian Federation, along with Belarus and Kazakhstan put a trade barrier in place in 2012; Taiwan, inspired by the EU Seal Regime, banned the trade in marine mammal products in 2013; Switzerland banned seal products in 2017; and India banned commercial seal products in 2018. Along with Mexico, which had already banned seal and other marine mammal products as a result of the North American Free Trade Agreement in 2006, important markets for seal products have now been closed.

Questions and Research Tasks

- How are seals being dealt with in your country (provided that there are seals where you live)?
- Can a principal opposition to the utilisation of a species be based on moral concerns?
- Does a blanket ban on seal products improve animal welfare in seal hunts? Discuss this issue with your colleagues, bearing in mind other options of regulating the seal trade.
- How can sealers be motivated to stick to high animal welfare standards?

[38]Sellheim (2016b).

References

Bailey TA (1935) The North Pacific Sealing Convention of 1911. Pac Hist Rev 4(1):1–14

Barry D (2005) Icy battlegrounds: Canada, the international fund for animal welfare, and the seal hunt. Breakwater Books Ltd, St John's

Busch BC (1985) The war against the seals. A history of the North American seal fishery. McGill-Queens University Press, Montreal

Cambou D (2013) The impact of the ban on seal products on the rights of indigenous peoples: a European issue. Yearb Pol Law V:389–415

COWI (2008) Assessment of the potential impact of a ban on trade in seal products. COWI, Copenhagen

EFSA (2007) Animal welfare aspects of the killing and skinning of seals– scientific opinion of the panel on animal health and welfare. EFSA, Parma

European Commission (2006) Communication from the Commission to the European Parliament and the Council on a Community action plan on the protection and welfare of animals, SEC (2006) 65, COM/2006/0013 final, 23 January 2006. European Commission, Brussels

European Parliament (1981) Report drawn up on behalf of the Committee on the Environment, Public Health and Consumer Protection on Community Trade in Seal Products and in Particular in Products Deriving from the Whitecoat Pups of Harp and Hooded Seals (Pagophilus groenlandicus and Cystophora cristata). PE 75.783/fin, 25 November 1981. European Parliament, Brussels

Frum B (1978) CBC Interview with Paul Watson. https://www.youtube.com/watch?v=H_gTBDFTXE0

Howse RL, Langlille J (2012) Permitting pluralism: the seal products dispute and why the WTO should permit trade restrictions justified by non-instrumental moral values. Yearb J Int Law 37 (2):367–432

Mossop J (2013) Marine mammals in the Antarctic treaty system. In: Molenaar EJ, Oude Elferink AG, Rothwell D (eds) The law of the sea and the polar regions. Interactions between global and regional regimes. Brill Nijhoff, Leiden, pp 267–292

Nielsen L, Calle MA (2013) Systemic implications of the EU—seal products case. Asian J WTO Int Health Policy 8:41–75

Sellheim N (2015) 'Direct and individual concern' for newfoundland's sealing industry? When a legal concept and empirical data collide. Yearb Pol Law VI:466–496

Sellheim N (2016a) Early sealing regimes: the Bering sea fur seal regime vis-à-vis Finnish–Soviet fishing and sealing agreements. Pol Rec 52(1):109–114

Sellheim N (2016b) The narrated 'other'—challenging Inuit sustainability through the European discourse on the seal hunt. In: Hossain K, Petrétei A (eds) Understanding the many faces of human security perspectives of Northern indigenous peoples. Brill Nijhoff, Leiden, pp 56–73

Wenzel G (1991) Animal rights, human rights. Ecology, economy and ideology in the Canadian Arctic. University of Toronto Press, Toronto

Further Reading

Mirovitskaya NS, Clark M, Purver RG (1993) North Pacific fur seals: regime formation as a means of resolving conflicts. In: Young O, Osherenko G (eds) Polar politics—creating international environmental regimes. Cornell University Press, New York, pp 22–55

Wegge N (2013) Politics between science, law and sentiments: explaining the European Union's ban on trade in seal products. Env Pol 22(2):255–273

The International Legal Framework for Polar Bears

7

7.1 Introduction

The polar bear (*Ursus maritimus*) is one of the largest bear species in the world and indeed an iconic species. It stands synonymous for the drastic effects of climate change on the Arctic environment and its almost hopeless battle against the impacts of humankind. In recent years, increasing horror stories about the polar bear have emerged in media discourse: the polar bears is on the brink of extinction; the polar bear is starving due to the effects of climate change; or the polar bear invades human settlements in search for food.

As we have seen in Chap. 2, the realities are far more complex. The Arctic Biodiversity Assessment (ABA), for instance, has found that of the 19 distinct polar bear populations all over the Arctic, seven are declining, four are stable, one is increasing, and the status of the remaining seven is unknown.[1] The claim of a declining polar bear is therefore imprecise. Similarly, the rather recently circulating image of a starving polar bear allegedly due to climate change caused high waves in social media. However, an extremely skinny polar bear in search for food can have numerous reasons and the causes for such physical state can also be related to disease or other factors. However, such emotional charging of the polar bear does certainly not miss its mark and attempts have been made to further protect this species from any human exploitation.

What is often missed in the discourse on polar bear protection is the fact that a regime has been in place since 1973 that puts the polar bear under protection by its range states: the Agreement on the Conservation of Polar Bears (ACPB),[2] which was concluded between Canada, the United States, Norway, Denmark/Greenland and the Soviet Union at the height of the Cold War. It entered into force in 1976 and has been in force since then.

[1]Meltofte (2013), p. 115.

[2]Agreement on the Conservation of Polar Bears, 15 Nov 1973 (13 ILM 13).

N. Sellheim, *International Marine Mammal Law*,
https://doi.org/10.1007/978-3-030-35268-4_7

7.2 The Origins of the Polar Bear Agreement

Concerns for the wellbeing and conservation of polar bears is not a phenomenon of recent times but has emerged already in the wake of the Second World War. During the war, the Arctic regions had become significantly easier accessible due to the leaps in technology—a development which would continue in the times after the war. This meant that apart from military personnel, also private citizens were now able to access the Arctic regions much better, putting Arctic living resources, including the polar bear, in much closer reach of humans. Of course, polar bears had been hunted for centuries by the Arctic's indigenous population, but the commercialisation of polar bear products, especially its hides, paired with high-powered rifles and other military technology put increasing pressure on polar bear populations. In addition it became clear that also indirect threats were becoming increasingly strong in the Arctic: pollution from non-Arctic sources as well as the first signs of a warming Arctic started to show their first effects.

During the 1960s it became clear that polar bear populations were declining to a degree that put their long-term survival at risk. At that time, however, Arctic cooperation in its contemporary form did not exist and research and scientific efforts on the polar bear were scattered. Also the bipolar geopolitical landscape made a collaborative scientific effort difficult. In spite of this, the United States took the initiative to set a process in motion that would lead to the conclusion of an agreement on scientific cooperation and ultimately conservation of the polar bear. The first steps were taken in September 1965 when the First Scientific Meeting on the Polar Bear was held in Fairbanks, Alaska. Representatives of the polar bear range state governments, research and management institutions attended the meeting along with the IUCN. Although the meeting could not agree on concerted research efforts, particularly because of the opposition from the Soviet Union, the final statement[3] nevertheless encouraged the range states to exchange scientific findings on the polar bear in order to be able to put a management regime in place. Until this would happen, the final statement encouraged each range state to take individual action it considered necessary to protect polar bears. While this was a rather general point of departure without much concrete action, we must bear in mind that the meeting comprised delegates from western states and the Soviet Union. From the outset, negotiations and exchange of information was difficult in these muddy geopolitical waters. It is ever more positively surprising that the meeting agreed upon a concrete course of action: the protection of polar bear cubs and females accompanied by cubs throughout the year.

A crucial role in the further advancement of information exchange was afforded to the IUCN. The IUCN has had a reputation of impartiality and its focus on conservation was considered non-political. Therefore, the first meeting suggested the IUCN to be the information hub of polar bear research and that polar bear range states were to submit their research findings to it. It was therefore the IUCN which

[3]PBSG (2009).

started the preparations for a Polar Bear Specialist Group (PBSG). The PBSG held its first official meeting in Morges, Switzerland, in 1968. Compared to the meeting of 1965 in Alaska, the first PBSG gathering was merely attended by 12 delegates, none of which were from Denmark (for reasons unknown). The meeting did not produce official proceedings since many of the issues discussed were considered confidential, once again reflecting the difficult geopolitical situation at hand. These referred particularly to issues relating to international waters and country delegates wanted to have open discussions without running into potential troubles afterwards.

One of the crucial disagreements in the meeting was the question of how to handle the polar bear population: is it scientifically justifiable to consider it as one circumpolar population? Or are there discrete individual populations? This question alone unveiled the significant lack of data on polar bears as well as different assessment methods, both of which would make the establishment of a management plan difficult. After all, it was particularly the Soviet Union which argued that polar bears should be considered as one circum-Arctic population. Agreement was reached on the finding, however, that the numbers of polar bears were going down. In the Soviet Union, polar bear hunts had been prohibited already since 1956 and its representatives also requested the other nations to put in place a total ban on polar bear hunts. Since in Canada, the US and Greenland polar bears had played an important socio-economic role for the indigenous populations, a total ban on polar bear hunting could not be achieved. But it was clear already during the 1968 meeting that some form of agreement had to be reached.

The first meeting of the PBSG was, however, a success in a different way it was agreed that regular meetings were supposed to take place every 2 years. In 1970, therefore, the PBSG came together again in order to further advance the cooperation on polar bear research. By then, bi- and multilateral research efforts had already been launched and polar bear research cooperation was ongoing between Canada, Norway, the US and Denmark. Although research cooperation with the Soviet Union did not prove to be successful, the 1968 meeting nevertheless succeeded in agreeing on the standardisation of research methods, which enabled the 1970 meeting to present streamlined results. Particularly the tagging technique brought clarity to the question of whether or not the polar bear population actually consists of several smaller populations: by 1970 it had become clear that there were at least five distinct polar bear populations, in Spitsbergen-Franz Josef Land-East Greenland region; in the Hudson Bay region in Canada; the high Canadian Arctic; the high Canada-eastern Alaska region; and the western Alaska-eastern Soviet Union region. With this finding, a main obstacle as to the nature of a potential polar bear agreement was out of the way.

The main difference between the first meeting in 1965 and the subsequent PBSG meeting was that the PBSG was operating under the IUCN and thus delegates were not under instructions from their respective national governments. This provided for a much more open and less geopolitically influenced stage for negotiations. Additionally, the polar bear specialists themselves had good relations with ministries and decision-makers on a governmental level, particularly in Norway and Canada. To advance the fruitful discussions and finally reach consensus on an agreement, the

PBSG decided to keep the meetings small and personal. In Norway and Canada, support from the ministries was strong due the personal connections to the PBSG. After the 1970 meeting, and with an eye on the subsequent meeting in 1972, IUCN had officially started working on a draft for an international polar bear agreement.

At the 1972 meeting, finally, the PBSG confirmed that all governments of the polar bear range states had informally voiced their support for a polar bear agreement. The main outcome of the 1972 meeting was therefore the call for the conclusion of an international agreement on the conservation of polar bears. But irrespective of the progress on this matter, the PBSG furthermore recommended a full ban on the hunting of polar bears in the high seas from 1973 onwards. As in the case of seals, pelagic hunting constituted a major threat to polar bear populations. This recommendation also included the hunt on pack ice, but excluded those hunts conducted by indigenous (local) populations. Interestingly, the recommendation refers to the "traditional rights of local people who depend on this [*sic*] resources."[4] Although neither 'traditional rights' nor 'local people' were more closely defined, the meeting took as a point of departure that a convention on polar bears was to take the socio-economic and cultural needs of local people into account.[5]

Yet another important aspect was considered: the wider ecosystem approach of polar bear management. The 1972 meeting also adopted a recommendation on the protection of ringed seals, which had been identified as the main food source of polar bears. In order to efficiently protect polar bears, also its main food source and its main hunting grounds had to be protected. The result was a recommendation for the advancement of studies and programmes on ringed seal management.[6] The 1972 meeting therefore ended with two recommendations which were rather far ahead of its time since, first, the rights of indigenous peoples had not been established as a principle in environmental decision-making and, second, that an ecosystem-based approach was included in the overall management scheme of polar bears, which the PBSG envisioned. At a second meeting, which took place during the IUCN General Assembly in August 1972 in Banff, Canada, the PBSG discussed a revised draft of the polar bear agreement. Although the meeting did not produce any minutes or other documentation, its results were informally communicated to the five Arctic governments. After all, neither the PBSG nor the IUCN could formally enter into the agreement and it was in the hands of the governments to decide on how to proceed.

Under the leadership of Norway, the draft provided by the IUCN was discussed in more detail. Two particular issues played a role in this regard: on the one hand, the recently adopted MMPA in the United States placed Canada in a difficult situation as regards its own delegations since terrestrial mammals (which polar bears were legally classified as) fell under the auspices of the territories and provinces and not under federal control—unless when dealt with under international agreements. All

[4] PBSG (1972a), Res#1-1972.

[5] PBSG (1972b), p. 14.

[6] PBSG (1972c) Res#4-1972.

other polar bear range states controlled polar bears federally, which meant that it should not have been the Canadian federal government but the provinces and territories negotiating the agreement. This was resolved by adding a polar bear scientist representing the territories and provinces to the Canadian delegation. The other and more significant event was the start of the third UN Conference on the Law of the Sea in 1973. The conference would last for 9 years and would result in the UNCLOS.[7] The political and legal sensitivity of the matters that were discussed by the conference also reflected onto the polar bear convention. Norway's overall approach was that the polar bear convention could serve as a stepping stone towards wider Arctic cooperation, particularly with regard to Arctic wildlife. Given that this would transcend national boundaries and affect issues relevant for the law of the sea conference, this was rather quickly dismissed. In the preparatory phase to the November 1973 meeting when the parties would get together in Oslo to discuss the polar bear agreement, a wider Arctic approach was therefore not chosen and the sole focus rested on the conservation of polar bears.

7.3 Finding Common Ground: The 1973 Polar Bear Agreement

Prior to the meeting, issues circling around high seas conservation were discussed and commented on. Terms such as "jurisdiction", "at sea beyond its territory" or "in the regions they [i.e. polar bears] inhabit" were legally too sensitive to be included in the agreement. Given that the Cold War was in full swing, finding common ground also concerning the rights of indigenous populations was difficult. After all, it was considered that indigenous populations should retain their right to hunt and sell polar bears and their hides. In order to do so, however, access to ports was necessary and given the difficult geopolitical situation—after all, the Arctic was a 'theatre' of the Cold War—access to ports by foreign vessels was a sensitive issue. Therefore, in the period before the meeting in November, it was yet unclear how the IUCN as the organisation in chart of drafting the agreement, was to incorporate this into the new draft under consideration.

The year 1973 saw also the negotiation, and ultimately adoption, of CITES in Washington DC. The United States, as one of those countries strongly pushing for a conservationist agenda, suggested to negotiate and agree upon an Interim Agreement on polar bears as a side agreement at the CITES meeting in January/February of 1973. Concerning the sovereignty issues involved in the polar bear agreement and fact that the Soviet Union had placed polar bears under strict protection since 1956, this suggestion was rejected. But as we will see, it is the United States which has unsuccessfully tried on several occasions to place the polar bear under a strict CITES umbrella.

As we have seen above, it has been particularly Norway which has signalled from the very early stages onward that it would take great interest in the adoption of a

[7] United Nations Convention on the Law of the Sea, 10 December 1982 (1833 UNTS 3).

polar bear agreement. It therefore informed the IUCN that it would be willing to host the crucial meeting for a polar bear agreement in Oslo. The IUCN, in turn, officially asked Norway to do so in early 1973. At this point, the agreement under discussion was considered under temporary terms, which means that it was rather an Interim Agreement than a full-scale, open-ended convention or other type of treaty. The draft that was to be discussed at the November meeting was therefore named Protocol on the Conservation of Polar Bears by the IUCN. Under international law, a protocol is an addition to an already existing legal instrument, which in the case of polar bears did not exist. In this sense, a protocol on the conservation of polar bears is legally speaking misleading and would not correspond to the legal realities. The IUCN aimed to avoid the term 'convention.' Given its legally-binding nature under international law, it would have been more appropriate, however. But given the sensitivities of international law-making between the two blocs of the Cold War, a set-in-stone convention was not considered suitable.

When the five Arctic nations finally met in Oslo, many issues had to be tackled. Once again it must be borne in mind that formal Cold War negotiations were not an easy task. Particular weight thus fell on informal negotiations in which especially the Soviet delegation was able to present their views without having to formally do so. This was especially the case since the Soviet delegates, in the formal setting, indicated that they did not have the authorisation to sign any agreement under negotiation. In informal settings, however, it was communicated to the other parties that also the Soviets were hoping for the adoption of an agreement.

The problem, still, was to find legal language that would not affect any negotiations under the law of the sea conference, which took place in parallel to the polar bear meeting. Particularly language referring to territorial waters and the high seas proved to be tricky. So the path for the conference to tread was to avoid these issues altogether. It was therefore agreed to prohibit all taking of polar bears, irrespective of where this taking was to occur. This meant that the agreed upon language broadly states that "The taking of polar bears shall be prohibited."[8] Five exceptions to that blanket prohibition were inserted: for scientific purposes; for conservation purposes; for marine management purposes to protect other living resources and provided that the products deriving therefrom were not placed on the market commercially; when polar bears are hunted by 'local people using traditional methods'; and "wherever polar bears have or might have been subject to taking by traditional means by its nationals."[9]

The latter two exceptions were inserted primarily at the behest of the United States and Canada as well as Denmark due to their respective Inuit population and their centuries-old interaction with polar bears. As mentioned above, the term 'indigenous' or 'native' is absent in the legal language applied, making it theoretically possible also for non-indigenous hunters to make use of this exception. The question why the terms were not used can be answered by looking at the Constitution

[8]Polar Bear Agreement, article I.

[9]Polar Bear Agreement, article III (e).

of Alaska.[10] Here, Article 8 outlines the provisions for the use of natural resources, which includes wildlife. No differentiation is made between indigenous or non-indigenous Alaskans and the principle of 'common use' is constitutionally enshrined. Breaking with this principle in the polar bear agreement would consequently be unconstitutional in Alaska. As a consequence, the language used refers to traditionality and not to ethnicity. In Svalbard, Norway, this has led to claims by Norwegians living in Svalbard to be considered under this exception. The Norwegian government, however, made clear that hunts conducted by temporary workers in weather stations or elsewhere in Svalbard would not be considered 'traditional' and consequently dismissed these claims.

With the trickiest legal language issues out of the way, the negotiations also focused on the nexus of management and polar bear research. Since a top-down approach was not the most aspired means to achieve the objective of conserving polar bears, it was agreed to somewhat scale down the provisions on research and management cooperation. While the agreement itself is legally binding and thus the language used is that of obligation ('shall'), the qualifiers such as 'as appropriate' or 'as may be necessary' essentially leaves it to the party to decide based on particular circumstances.[11]

An important feature of the Polar Bear Agreement is Article II. The article is in principle an advancement of the earlier resolution on the protection of seals although it does not mention seals at all. Instead, the article requires the parties to protect "the ecosystems of which polar bears are a part," making the ecosystem approach a legally binding principle in an international agreement. As an international norm the ecosystem approach would emerge only as part of the 1992 United Nations Conference on Environment and Development (UNCED) in Rio de Janeiro (Rio Conference). Ultimately, the ecosystem approach would find its way into the Convention on Biological Diversity (CBD)[12] and many subsequent agreements. The Polar Bear Agreement was therefore a forerunner in this regard and underlines its progressiveness.

The Agreement for the Conservation of Polar Bears was signed in its present form on 15 November 1973 by all states except for the Soviet Union since the Soviets did not have authorisation to sign any agreement. One of the requirements for the agreement to enter into force was the ratification of three parties. Canada ratified the agreement in 1974, followed by Norway in 1975. With the Soviet Union's signature and immediate ratification in 1976, the Polar Bear Agreement entered into force. The United States ratified it in the same year while Denmark was the last one to do so in 1978. To underline Arctic sovereignty, the Polar Bear Agreement was only open for signature for the five polar bear range states. Other countries, which might also be involved in the hunt for polar bears were excluded from the agreement.

[10]Constitution of the State of Alaska, 24 April 1956.

[11]See for example Polar Bear Agreement, article VI, VII, VIII.

[12]Convention on Biological Diversity, 5 June 1992 (1760 UNTS 79).

Although the parties wished to have a temporary term for the agreement removed, as we discussed above, initially it was to be in force only for five years. That meant, by 1981 the parties had to decide whether or not the agreement was to continue. In January 1982, the parties therefore met in Oslo to discuss its future. Norway, inspired by article II of the agreement, once again sought to expand the purview of the agreement to include other Arctic species. It therefore aimed to make the Polar Bear Agreement the starting point for wider Arctic environmental cooperation. Both the United States and Canada initially agreed that this might be an option, but in the end show little interest to expand the agreement's scope. Based on the tense geopolitical environment that the Arctic represented, the Soviet Union did not want a meddling in its Arctic affairs and rejected this effort altogether. Consequently, no amendments or expansions were undertaken at that meeting. While that was the case, the parties agreed for the agreement to be in force indefinitely.

Wider Arctic environmental cooperation was not taken up by the parties. The Cold War prevented effective cooperation in this regard. But the Polar Bear Agreement shows that despite wider geopolitical tensions, Arctic environmental cooperation is possible. The agreement was furthermore adopted in a precautionary manner, meaning that polar bear populations were indeed in decline, but, contrary to other species, not decimated upon the agreement's adoption. Throughout the 1980s the static borders between East and West slowly started to erode. For the Arctic, the 1987 speech by Soviet Secretary General Mikhail Gorbachev in Murmansk marked the beginning of a new era of cooperation. He envisioned the North Pole to be a 'Pole of Peace' and the proposals he put forth in his speech included scientific exchange and environmental exchange in the Arctic. As a result, Finland initiated the Arctic Environmental Protection Strategy (AEPS) in 1992, which would transform into the still-existing Arctic Council in 1996. Apart from the five polar bear range states, also Finland, Sweden and Iceland as well as six Arctic indigenous organisations are included in the working procedure of the Council.[13] Up until the time of writing in the summer 2019, three legally binding agreements have been concluded under the umbrella of the Arctic Council: on search and rescue (2011); on oil pollution preparedness and response (2013); and on scientific cooperation (2017).

7.4 Polar Bear Management Agreement in the Southern Beaufort Sea

With the Polar Bear Agreement in place, the protection of polar bears was not a finalised endeavour. This soon became evident in the Southern Beaufort Sea, the marine area between Northern Alaska and the Northwest Territories in Canada. Here it was assumed that the abundant polar bear populations belonged to different sub-populations and thus could be managed independently in the US and in Canada. Throughout the 1980s it became clear, however, that the population was

[13]Keskitalo (2004).

one transboundary population, requiring joint management schemes on both the US and Canadian side of the border. We must remember that in the United States and in Canada two significantly different management schemes and legal regimes existed. In the US, the Marine Mammal Protection Act of 1972 prohibited the hunting of all marine mammals except for the native population, as long as the use of these products was for their own use. The sale of marine mammal products, in this case polar bear hides, meat or bones to third parties was strictly prohibited. Notably though, the MMPA does neither set quotas nor determines the time of the year when the aboriginal polar bear hunt is to occur. Only in case of a depletion of a marine mammal population the federal government is able to put strict limitations also on the aboriginal subsistence hunt.

On the Canadian side, the northern Northwest Territories towards northern Yukon is the homeland of the Inuvialuit, an Inuit sub-group with its own language and culture. Already in 1968 a quota system was introduced that ascribed a specific polar bear quota to each community in the region of the Southern Beaufort Sea. Since 1974 up to 1984, the government of the Northwest Territories undertook several specific efforts to further protect the polar bear from harvesting, for instance the prohibition of harvesting bears in dens or the prohibition of hunting bears in the fall. This did not mean that the commercial sale of polar bear products was prohibited and Inuvialuit hunters were therefore able to place their produce on the commercial markets. Also the transfer of the polar bear quotas to accommodate outside sport hunters was possible. In 1984, the Inuvialuit Final Agreement (IFA)[14] was implemented. The IFA formalised the ownership of the Inuvialuit of 91,000 km^2 of land in the Southern Beaufort Sea and established co-management procedures of both the Inuvialuit and the federal government in several councils and committees. This means that decisions on wildlife management are taken by both the Inuvialuit and the federal government through equal representation in management bodies. Under the IFA, the Inuvialuit have the prioritised right to harvest marine mammals and to transport marine mammal products across the border to the Yukon Territory. They are entitled to the sale, trade and bartering of marine mammal products to other Inuvialuit in both states.

As we can see, the legal environment between Alaska and the Northwest Territories was significantly different, but since the polar bear population was shared, a common regulatory mechanism was necessary. In 1985 the initiative was taken up by the Alaskan municipality of the North Slope Borough (NSB), primarily comprising Iñupiat, which had been established in 1972, and the Canadian Inuvialuit Game Council (IGC). The point of view was that formal, long-lasting governmental negotiations between the US and Canadian governments would possibly undermine the principle of good stewardship and effective polar bear conservation. In September 1986, a Memorandum of Understanding was signed, which established a joint commission and a technical committee. The joint commission was tasked with the development of a management plan for polar bears in the Southern Beaufort

[14]Inuvialuit Final Agreement, 5 June 1984.

Sea. The first draft agreement was presented in March 1987 and both sides agreed on its wider terms. On 29 January 1988 this agreement was formalised in the Polar Bear Management Agreement for the Southern Beaufort Sea.[15]

The uniqueness of the 1988 management agreement lies in the fact that it was not negotiated by governments, nor was it initiated by government authorities. Instead, it was the resource users which took up the initiative to enter into an agreement to ensure the efficacy of the 1973 Polar Bear Agreement and to make polar bear conservation feasible. Although it was not legally enforceable and constituted merely a 'gentlemen's agreement' between the NSB and the IGC, it is nevertheless based on existing law, i.e. the ACPB and on Canadian legislation, as well as on scientific findings. However, on the Canadian side, the agreement was somewhat enforceable since the Joint Commission's main role was the setting of quotas. These, naturally, did not exceed those under Canadian regulations. Although the agreement did not carry legal weight in the traditional sense, it set a precedent for other user-based resource management regimes. The most important development in this regard was the acceptance of the quotas set by the Joint Commission by the Canadian hunters and trappers committees, which could be found all over the Northwest Territories. These committees were involved in the quota negotiations with the federal government and once they accepted the advice by the Joint Commission as regards quotas, its authority was underlined.

The 1988 agreement remained unchanged until the year 2000. All in all, the agreement was considered a success and when the two bodies met again for an evaluation and possible extension, it became clear that both sides remained deeply committed to it.[16] While the main body of the agreement remained as in 1988, one important addition was made: bears that are killed while threatening human life or safety as well as bears that are killed for research purposes should be counted as part of the overall quota. This provision added an extra layer of conservation efforts to the agreement which the 1988 version did not entail.

The Inuvialuit-Iñupiat polar bear management agreement remains in place up to the point of writing and underwent its last slight revisions in 2011. A key feature of the joint management between the NSB and the Inuvialuit is that of traditional knowledge, *Inuit Qaujimajatuqangit*, paired with scientific knowledge. This means that it is not just scientific elements that dictate the quota, but socio-cultural considerations play a significant role as well. The same accounts for the Polar Bear Management Agreement for the North Beaufort Sea and Viscount—Melville Sound Polar Bear Populations between the Inuit of the Kitikmeot West Region in Nunavut and the Inuvialuit, which was concluded in 2006. Since this agreement is a domestic agreement in Canada, we will not delve deeper into it. In 2008, however, the Inuvialuit-Iñupiat Agreement, which was indeed user-driven, also found its reflection on the federal level when the United States and Canada entered into an

[15] Polar Bear Management Agreement for the Southern Beaufort Sea, 29 January 1988.

[16] Brower et al. (2002).

MoU for the conservation and management of shared polar bear populations.[17] The MoU coincided with the US listing of the polar bear as threatened under its Endangered Species Act and very broadly aims for coordinated efforts for the conservation and management of joint polar bear populations. The MoU highlights the inclusion and exchange of traditional knowledge and science regarding polar bear use and management. The MoU aims for the establishment of a Bilateral Oversight Group, but up to this point it remains unclear whether or how this has been put into practice. It makes unmistakably clear that it is not legally binding and thus does not put any legal obligations on neither Canada or the United States. It is therefore merely an expression of intent rather than a bilateral legal instrument.

7.5 Managing Polar Bears in Kane Basin and Baffin Bay

Just one year after the MoU between Canada and the United States, the former entered into another MoU on polar bear management. This time the MoU was concluded between the governments of Canada, of Nunavut and of Greenland,[18] concerning shared polar bear populations in Baffin Bay and Kane Basin, the waters between Ellesmere Island in Canada and Greenland. The issue of over-harvesting had been known for several years. As the basis for the agreement which was the response to unsustainable harvest levels serves the ACPB. While the MoU is not legally binding itself, the existing legal regimes for indigenous peoples and associated harvesting rights as well as the land claims under the legal frameworks in Nunavut and Greenland create legal obligations, as the MoU recognises in its Preamble.

The overall purpose of the MoU is the conservation and sustainable management of polar bears. But in the spirit of the ecosystem approach, the MoU approaches polar bear conservation through the protection of ecosystems and natural habitats while also including the restoration and revitalisation of depleted populations as well as sustainable utilisation.[19] To advance the overall purposes of the MoU, a Joint Commission was established, whose primary task it is to make recommendations on sustainable harvest limits. A particular feature in this regard is that this recommendation is not communicated to the management bodies directly, but to a User-to-User Working Group, which is formed under the commission, which itself makes recommendations back to the commission in how far the harvest levels are allocated between Greenland and Canada. Once the commission finds the recommendations

[17]Memorandum of Understanding between Environment and Climate Change Canada and the United States Department of the Interior for the Conservation and Management of Shared Polar Bear Populations, 2008.

[18]Memorandum of Understanding between the Government of Canada, the Government of Nunavut, and the Government of Greenland for the Conservation and Management of Polar Bear Populations, 31 October 2009.

[19]Ibid, Art 2.

by the Working Group acceptable, it communicates these back to the relevant governmental bodies.

Apart from recommending harvest levels, the Joint Commission furthermore serves as the hub for harvest data, scientific activities and traditional knowledge surrounding polar bears in the region of applicability. The Joint Commission comprises eight members: one from the government of Canada; three from Nunavut (one governmental, one wildlife manager, one business); and four from Greenland (two governmental, one wildlife manager, one hunter organisation). *De facto* it is therefore an indigenous co-management body.

7.6 Managing Polar Bears in Alaska and Chukotka

The Canadian Inuvialuit are not the only group having entered into an agreement with the Alaskan side. After all, there are polar bear populations that are shared between Alaska and Chukotka in the Russian Federation. After the end of the Cold War and with the increase in Arctic cooperation facilitated by the AEPS, in 1995 US and Russian officials concerned with polar bears began meeting in order to develop a joint management plan. Several US agencies and their Russian counterparts were involved in these meetings. By 1997 the US Fish and Wildlife Service, a key player in the meetings, had finalised an environmental assessment for the region and formally requested the State Department for the authorisation to start negotiations for a bilateral agreement on polar bear conservation with Russia.

The agreement was negotiated between January 1998 and finalised in March 2000. On 16 October 2000, the US and Russia signed the Agreement on the Conservation and Management of the Alaska-Chukotka Polar Bear Population.[20] This legally-binding agreement entered into force in September 2007 and advances the terms of the ACPB in the Bering, Chukchi and eastern East Siberian Seas. While in principle a regional version of the Polar Bear Agreement, it nevertheless differs in some way. First and foremost, the Alaska-Chukotka Agreement is fundamentally more geared towards the inclusion of native peoples. Throughout the Agreement the role of native peoples is acknowledged and reaffirmed and already the Preamble highlights the "essential role of the native people of Alaska and Chukotka in the conservation of the Alaska-Chukotka population of polar bears."[21] Second, in 1994 the US amended the Marine Mammal Protection Act to allow the importation of polar bear hides from qualified sport hunts in Canada. This marked a significant step towards indigenous self-determination as to how to use their quotas and what to do with the yields. When the ACPB was concluded in 1973, this amendment had not been in place and a total ban on any imports of marine mammal products existed. Therefore, the 2000 Alaska-Chukotka Agreement explicitly states that it is aimed *not*

[20]Agreement on the Conservation and Management of the Alaska-Chukotka Polar Bear Population, 16 October 2000.

[21]Ibid., Preamble.

to authorise commercial trade in polar bears, but equally *not* to "limit the ability of native people [. . .] to create, sell, and use traditional articles associated with native harvest of polar bears."[22]

The primacy of native involvement is furthermore underlined in the US-Russia Polar Bear Commission, which is established under the Agreement. The four-member commission consists of two members of each country, one of which is to be an Alaskan and Chukotkan native. A rather long catalogue of tasks is allocated to the Joint Commission. Broadly summarised, the Joint Commission facilitates the communication and cooperation between governmental bodies and native peoples on both sides of the border concerning the conservation and use of polar bears. The Commission furthermore sets the annual sustainable harvest level and associated limits for the annual take. This occurs not only based on scientific data, but takes equally into account the traditional knowledge of native peoples. Although the subsistence use of polar bears is explicitly fostered in the agreement, the Commission has the mandate to limit subsistence takes of polar bears in case of conservation concerns.

Relatively little information is available on both the history and the functioning of both the agreement and the Joint Commission. However, apart from the ACPB, the Alaska-Chukotka Polar Bear Agreement is the only legally-binding bilateral polar bear agreement. The overall population status of the Alaska-Chukotka polar bear population is unknown. While it is likely that illegal takes by the local population have dropped since the 1990s, it can be assumed that the combination of legal and illegal takes, paired with other threats may put this subpopulation under pressure. The increased presence of polar bears on land may point to increasing impacts of climate change. However, it also appears that polar bears have, in spite of decreasing sea ice, not changed their habitat preferences, which might indicate that the Alaska-Chuktokan polar bear population might be increasing,[23] possibly also as a result from this agreement.

7.7 Uplisting the Polar Bear Under CITES

When the Polar Bear Agreement was concluded in 1973, it was not the only international agreement that year which would consider the conservation status of polar bears. In 1973, CITES was adopted, which we already got to know in Chap. 4. To recapitulate, CITES works to protect species from unsustainable trade and has therefore put in place three Appendices, based on which trade is regulated. Species listed on Appendix I cannot be traded in, unless under very exceptional circumstances. Species listed under Appendix II can be used for limited international trade; Species listed under Appendix III may be subject to national trade measures and further international cooperation is needed to avoid unsustainable trade.

[22]Ibid., Art VII.1.

[23]PBSB (undated).

Given that polar bears were, at the time of CITES' conclusion, considered to be under threat from overexploitation, it is no surprise that the negotiation of CITES also considered polar bears from the very beginning. When CITES entered into force on 1 July 1975, polar bears were one of the first species to be listed on CITES Appendix II. This meant that apart from the ACPB, a second regime, this time geographically much more far-reaching than the Polar Bear Agreement, was in place that would regulate the international trade in polar bear products. Since the ACPB is not open to other parties than the five polar bear range states, the CITES listing made the trade conducted by non-ACPB members significantly more difficult. How large the trade by non-range states is, is difficult to assess. John E. Scanlon, former Secretary General of CITES, noted in 2013 that "[a]round 80% of all such exports [of polar bear products] originate from one range State."[24]

Be that as it may, official CITES documentation does not make reference to the polar bear up until the 15th Conference of the Parties (CoP) in 2010. Here, the United States tabled a proposal to uplist the polar bear from Appendix II to Appendix I. The aim was to ban international trade in polar bear products altogether. This would have meant that the cross-border trade which had been established between Canada and the United States as well as the Russian Federation and the United States would have come to an end. The proposal was based on the domestic US Endangered Species Act under which in May 2008 the polar bear had been listed as 'threatened'. This listing meant that it was listed as being closer to reach the status of 'endangered' and thus under threat of extinction. Once listed, any other lethal take of polar bears apart from aboriginal subsistence hunts would be illegal. This also included cross-border movement of polar bear products.

The listing as 'threatened' did not make reference to international trade as one of the threats to the species. Instead, climate change and associated habitat loss was named as the primary cause for this listing. Interestingly, Interior Secretary Dirk Kempthorne, the Minister responsible for making such decision, noted that the Endangered Species Act would "not be used as a tool for trying to regulate the greenhouse gas emissions blamed for creating climate change."[25] Also when the US presented the proposal at the CoP in 2010, habitat loss and not international trade was named as the primary cause for polar bear population decline. However, the US was concerned that if polar bears continued to be listed on Appendix II, international trade as well as the sports and trophy hunts, particularly in the Russian Federation, may have a detrimental effect on the polar bear population.

Although some countries and NGOs were in support of the proposal, not surprisingly, Canada, Greenland and Norway, supported by Iceland and the European Union, opposed the proposal, stating that the trade in polar bear products was not market-driven, but based on adaptive co-management systems fulfilling subsistence needs. Reference was, of course, indirectly made to the different polar bear agreements and MoUs in place. Additionally, both Greenland and Canada had put

[24]Scanlon (2013).

[25]Greenemaier (2008).

in place export bans for specific polar bear populations in order to provide for sustainable population levels. These contexts, paired with the fact that the polar bear populations by far exceed 5000 individuals and therefore cannot be considered 'very small'—a precondition for Appendix I-listing—consequently did not correspond to the criteria for inclusion of species on Appendix I. When the proposal went for a vote, it was defeated by 48 parties in favour, 62 against and 11 abstentions.

At the following CoP16 in 2013, the United States took a second attempt to uplist the polar bear. Essentially using the same criteria, it now argued that a listing on Appendix I would reduce the overall pressure on polar bears. The discussions that ensued mirrored those from the previous CoP. The European Union proposed several amendments to the US proposal, aiming to find a consensus amongst the parties. These proposed amendments included, amongst others, the limitation of trade to polar bear range states only, different means to trace polar bears, and to take action against climate change in general. These proposed amendments were rejected by the CoP. When the original proposal by the US went to a vote, it was also rejected 38 in favour, 42 against and 46 abstentions.

For the time being this has been the last attempt by a CITES party to uplist the polar bear to Appendix I. With the ongoing decrease in Arctic sea ice, however, it may not have been the last. Whether the ongoing trade in polar bear products is indeed a significant factor for the protection of the species remains questionable.

7.8 Summary and Conclusion

Polar bears have found international attention as a powerful symbol of climate change, both in the Arctic and elsewhere. The starving polar bear has become a monument that transports a message of vulnerability and supports the discourse on the dying-out species. As we have discussed in Chap. 3, the conservation status of polar bears is far from clear. Moreover, what this discourse furthermore neglects is the fact that international and regional cooperative regimes exist that aim to protect, manage and utilise the polar bear.

The ACPB is indeed an important agreement both from a conservation perspective as well as from a perspective of geopolitics. It shows how in the Arctic the Arctic states have already managed during the Cold War to get together and work towards a common goal. In this case, the conservation of polar bears. But what the agreement also shows is that legal instruments that are drafted and ultimately implement by the nation states do not necessarily have to follow a top-down approach, but can also take into account the interests of the local population. Still, as the discussions around the uplisting of the polar bear from CITES Appendix II to Appendix I have shown, also this cooperation is not immune from top-down challenges. The example shows that although the US has been a party to the polar bear agreement and although local voices were heard at the meeting, this does not automatically translate into a more inclusive form of decision-making.

In order to alleviate problems related to polar bear management, the three agreements to manage polar bears in the North American Arctic are inherently inclusive since they are considered user-to-user agreements that take into account the views and practices of polar bear hunting communities. The fact that two are in any way legally non-binding may be seen as a weakness of these agreements, but as the other example in the Arctic of a non-binding form of cooperation, the Arctic Council, shows, legal non-bindingness may raise the legitimacy amongst Arctic peoples especially when the decisions that are made take into account their interests.

Unfortunately, I have yet to find recent studies (or even studies in the first place) that trace the efficacy and effectiveness of the different polar bear agreements. Therefore I am unable to comment on whether polar bear conservation has benefited from them. But irrespective of this, we can conclude that any cooperative agreement is better than no agreement at all.

Questions and Research Tasks

- How effective have the polar bear agreements been? Have polar bear populations benefitted from their conclusion?
- Does the way polar bear management has been approached serve as a 'blue-print' for other species elsewhere in the world?
- If it turns out that all polar bear populations are after all declining, should the polar bear agreement and the different management agreements be voided or amended to a degree that makes polar bear utilisation no longer possible?

References

Brower CD, Carpenter A, Branigan ML, Calvert W, Evans T, Fischbach AS, Nagy JA, Schliebe S, Stirling I (2002) The polar bear management agreement for the Southern Beaufort Sea: an evaluation of the first ten years of a unique conservation agreement. Arctic 55(4):362–372

Greenemaier L (2008, May 14) U.S. Protects Polar Bears Under Endangered Species Act. Scientific American. https://www.scientificamerican.com/article/polar-bears-threatened/. Accessed 16 May 2019

Keskitalo ECH (2004) Negotiating the Arctic. The construction of an international region. Routledge, New York

Meltofte H (ed) (2013) Arctic biodiversity assessment. Status and trends in Arctic biodiversity. Conservation of Arctic Flora and Fauna, Akureyri

PBSG (1972a) Res#1-1972: Protection of polar bears on the high seas. http://pbsg.npolar.no/en/meetings/resolutions/03.html#1-1972. Accessed 8 May 2019

PBSG (1972b) Proceedings of the Third Working Meeting of the Polar Bear Specialist Group, Supplementary Paper No 35. http://pbsg.npolar.no/export/sites/pbsg/en/docs/PBSG03proc.pdf. Accessed 8 May 2019

PBSG (1972c) Res#4-1972: Seal protection and research. http://pbsg.npolar.no/en/meetings/resolutions/03.html#1-1972. Accessed 8 May 2019

PBSG (2009) 1st Scientific Meeting on the Polar Bear, Fairbanks, Alaska 1965, Statement of Accord Approved by the Delegates. http://pbsg.npolar.no/en/meetings/press-releases/00-Fairbanks.html. Accessed 8 May 2019

PBSB (Undated) Chukchi Sea (CS). http://pbsg.npolar.no/en/status/populations/chukchi-sea.html. Accessed 16 May 2019

Scanlon JE (2013) Keynote Address, International Forum on Conservation of Polar Bears and Jubilee Meeting of the Parties to the 1973 Agreement on the Conservation of Polar Bears, Moscow, Russian Federation, 4–6 December 2013, https://www.cites.org/eng/news/sg/2013/20131204_polar-bear.php. Accessed 16 May 2019

Further Reading

Larsen TS, Stirling I (2009) The agreement on the conservation of polar bears – its history and future. Norsk Polarinstitutt, Tromsø

8 The Characteristics of International Marine Mammal Law

8.1 Introduction

The international legal landscape for marine mammals, which I have coined 'international marine mammal law', is wide and scattered. In the Introduction to this book I have argued that the main body of international law that international marine mammal law belongs to, is international environmental law as well as the law of the sea. For the most part, this statement still holds true, but we have seen that also other branches of (international) law comprise international marine mammal law.

From the early days of international, multilateral agreements relating to marine mammals, these agreements were driven by the incentive to exploit them. From the offset, it was seals which saw the first regulatory regimes. These regimes were not global, but focused on specific regions or seal populations. The Jan Mayen Treaty of 1874 paved the way in this regard as it was the first treaty to link the interests of different countries under one regulatory umbrella. While motivated by nations to further advance their seal hunts at Jan Mayen, the treaty set an important precedent for the future of marine mammal conservation and regulation for at least 100 years. This is to say that it corresponded to the state of knowledge of that time and that the ecosystem approach or other modern elements of environmental management did not exist at that time.

8.2 Geographic Limitations

Instead of addressing the wider marine ecosystem, the Jan Mayen Treaty was limited in its scope to one specific destination. These geographic limitations have become a feature of many, if not most, regimes that constitute international marine mammal law up to the point of writing. We must note, however, that these geographic limitations apply mostly to those regimes that are directly concerned with the management, or regulation, of marine mammals and marine mammal utilisation. The Bering Sea Fur Sea Regime, for example, held clear provisions as to the

N. Sellheim, *International Marine Mammal Law*,
https://doi.org/10.1007/978-3-030-35268-4_8

geographic scope of its provisions. It was therefore not necessarily the species of seal that was to be protected, or managed, but rather the seal species at a given location.

Another prominent example in this regard is the Schedule to the ICRW. Leaving aside the moratorium for a minute, the original idea of the Schedule contained geographic limitations for specific species. This means that the overall population of either one species of whales or either the super-family of cetaceans has been subdivided into different stocks that can be found within certain geographic boundaries. Depending on the species these are further subdivided into stocks that can, or cannot, sustain a certain degree of exploitation. Based on this, the amendments to the Schedule over the years have applied different criteria for exploitation on different stocks.

The idea of geographic limitations has furthermore been advanced in the case of whale sanctuaries. The establishment of whale sanctuaries is an effective tool of advancing the conservationist agenda of the International Whaling Commission. While the term 'sanctuary' implies that whales are free from pursuit, this does not mean that whales are free from other dangers. In other words, whale sanctuaries are merely attempts to limit—or prohibit—the hunt for whales in a specific area. This limitation, however, does not include dangers related to pollution, noise pollution, ship strikes or habitat degradation. The imposition of a zero catch quota for specific regions is consequently merely a very limited attempt to ensure the protection of whales, either for a specific region or in the world. As we have discussed in Sect. 4.4, whale sanctuaries can therefore not be considered marine protected areas in the strictest sense. I would consequently argue that marine protected areas that take into account a diverse threat-landscape within a given location, including ways and means to include human use of the marine environment, can be considered marine mammal sanctuaries. After all, by developing comprehensive strategies to protect the marine and coastal environment, for instance through coastal zoning which clearly denotes where vessels can operate and where they can't, also marine mammals are positively affected.

Geographic limitations, especially concerning bi- or multilateral environmental agreements, are, by definition, a necessity that take into consideration different sovereign territories of nation states. A regional agreement between two states can therefore not be applicable to a third state that has not signed or ratified an agreement (i.e. the *pacta tertiis* principle). The CMS has taken special measures to address this issue in the form of its Agreements and the different MoUs. Range states of a certain species that have neither become party to the convention itself nor to an Agreement are still considered in the organisational set up of the respective instrument. Contrary to the notion of whale sanctuaries, the limitations of which are artificially and politically determined, the limitations under the CMS and its agreements correspond to the limits determined by ecological parameters. The comprehensive approach of the CMS is inherently species-based, but is not limited to merely political boundaries or to the eradication of one particular threat. The OSPAR regime, on the other hand, is a regime which is also limited in its geographical scope and which also pays regard to marine mammals. While this consideration is also done on a species basis, it considers them as part of the overall marine environment in the North Atlantic. It therefore does not explicitly focus on marine mammals, but takes into account the latest scientific data on different population statuses and threats. From the outset, OSPAR was not designed as a regime for the protection of marine mammals, but as a regime that, within its geographical limitations, aims to tackle different sources of pollution in the convention area.

Geographical limitations within international marine mammal law thus follow a twofold approach. Older regimes that have a distinctly species-based set-up, particularly the sealing and whaling regimes, make use of geographical limitations either based on regional exploitation (e.g. the Bering Sea regime) or based on political stances on marine mammal exploitation (e.g. the ICRW and its whale sanctuaries). Whether these regimes translate into long-term protection of the species in question remains doubtful given the very limited umbrella of protection they afford.

The polar bear regime is somewhat different in this regard. Naturally, it is geographically restricted since polar bear habitat can only be found in the Arctic. The polar bear regime therefore follows the habitat of the bear instead of being artificially designated. The inclusion of utilisers of the species also points towards the fact that geographical limitation is not rooted in political decisions but rather corresponds to the realities on the ground. The conclusion of three additional agreements and the establishment of commissions that display co-management characteristics, i.e. governmental and non-governmental decision-making, makes this regime somewhat unique. Especially since it can be considered a type of hybrid regime: on the one hand it focuses on the polar bear explicitly and is therefore species-based; on the other, it takes polar bear habitat and resource users into consideration in the setting of quotas and modes of utilisation. Given that the polar bear agreement stems from the early 1970s, this is rather unique. After all, at that time, international discourse was on the verge of a shift towards a non-use agenda of marine mammals. Arguably, the polar bear was not part of this discourse since its representativeness for the negative effects of climate change had not been known. Only in recent times, international discourse has also started to consider polar bears as a species that is not to be used for human consumption. However, while that may be so, the overall legal landscape for polar bears has remained considerably unaffected and the polar bear agreement remains in force.

8.3 State Behaviour

International marine mammal law is deeply marked by the behaviour of different states or groups of states that stand representative for the way humans can interact with marine mammals. At the same time, it is also marked by successful cooperation, working towards the common goal of their protection.

From the very beginning of modern international law, the United States, along with other European states, has played a crucial role in its development. This is particularly true for the development of international marine mammal law. Already prior to the Second World War the United States pushed for more conservation-based approaches in whale management while other states that also had an interest in whale conservation considered whale management more through a lens of benefit for humans. Over the course of time, the US has developed to become an outspoken anti-whaling and anti-sealing nation, best exemplified by the Marine Mammal Protection Act. This being said, 'anti-sealing' and 'anti-whaling' are to be considered with a hint of caution: for the United States, it is not the seal or whale hunt *per se* which it opposes, but rather the hunts for commercial purposes. This is quite interesting since the United States has always pushed, and even significantly shaped, the free trade doctrine that has, in turn, been fundamental for the global economic system. In the case of marine mammals, the US displays a stringent precautionary approach: it is better not to hunt them in case of scientific uncertainty. This is further paired with the inherently emotional view on marine mammal hunts. On the surface, therefore, the US is a steadfast anti-use state.

When we dig a little deeper, however, we can see how contradictory this behaviour is. For instance, while the MMPA is in force, this has not prevented the US military from experimenting with dolphins or from keeping them, and other species, in captivity for human entertainment. Furthermore, the precautionary approach is not enshrined in US behaviour in other fora. For example, when the European Union banned certain meat products from the US that contained certain hormones, it was the United States that challenged this ban before the World Trade Organization in 1996. Here, the US argued that there is no proof that these hormones cause health problems and that it should be the European Union to provide scientific evidence for this harm. In other words, while the EU applied the precautionary approach, it was the US which challenged this.[1] Why the precautionary approach is applied in the context of marine mammals in US policy-making remains in the realm of speculation. It seems fair to assume that public opinion plays an important role, making this behaviour consequently inherently political. In addition, commercial trade in marine mammal products has not played an important role in the US economic system. Therefore, it is rather 'easy' for the United States to take this no-use approach—an approach which has also led the US to try to uplist the polar bear under CITES. Instead of tackling the long-term effects that climate change causes to polar bears, the US rather argues along lines that do not necessitate significant changes in the US economy.

Along with the United States go a rather large group of industrialised and developing nations that have taken up this view. The European Union's anti-sealing and anti-whaling policies do not cast doubt on this. Whether or not this corresponds to the moral view of the European public remains doubtful, however. In the case of sealing there is considerable evidence that shows that also within the European Union no such general moral view exists in the first place. I have therefore argued elsewhere that this opposition can rather be found in 'not wanting it' than in a

[1] See WTO (1996).

deeply-rooted moral standard.[2] Again, this would consequently stem from political rather than moral or even ecological or animal welfare considerations.

An interesting case in point is the case of Brazil, which has taken up a strict no-use stance within the IWC—best exemplified by the recent Brazil-led IWC Florianópolis Declaration of 2018. Brazil was one of the first states to join the IWC when it ratified the ICRW in 1950. It subsequently left the Commission in 1966 and then rejoined in 1974. Prior to its leaving the IWC, the country was a rather inactive member and on several occasions did not even join the meetings. Upon leaving the IWC, Brazil started to engage in large-scale minke whaling in the South Atlantic, killing several hundred whales per year, despite not having whaled for several hundred years. Only in the 1970s the country started to take an interest in environmental issues. Up to then, Brazil was considered a 'pirate whaling' nation that did not follow international rules. After rejoining the IWC, Brazil worked hard to rid itself of the 'pirate' stigma and started to develop an outspoken anti-whaling policy, for instance by promoting whale sanctuaries as a conservation tool. Yet, only from the 1990s onwards, it started to fully align itself with anti-whaling countries. At the 2018 meeting, Brazil's anti-whaling stance also translated in abstaining from a vote on the new ASW quota. Whether this means that in the future it will also principally oppose ASW remains to be seen. At the meeting, however, it became clear that Brazil is taking a lead in the so-called South American 'Buenos Aires Group' and increasingly pushes towards more conservation of whales. At the same time, Brazil advances the idea of non-lethal use of whales in the form of whale watching. Not surprisingly, the proposed South Atlantic Whale Sanctuary is very much in Brazil's interest.

On the other side of the spectrum we find countries such as Japan, Canada, Norway or Iceland. As has become clear in the foregoing chapters, Japan has pushed the pro-whaling agenda for many decades. But beyond that, Japan has always had a pro-use agenda towards marine living resources. Its involvement in the Bering Sea fur seal hunts and its long-standing history of whaling both in its coastal regions and in the Pacific and Southern Oceans stands exemplary in this regard.

Japan's overall science-based approach has marked its contributions to the IWC, justifying its commercial hunt for whales. At the same time, it has also supported conservation efforts, yet as long as its own scientific findings were in support of such measures. However, contrary to other countries, Japan's overall approach to the IWC has been marked by its own national interest. The attempted establishment of the 'small-type coastal whaling' category in support of its coastal minke whaling activities are but one example. A blanket ban on commercial whaling was therefore opposing Japan's IWC policy in a two-fold manner: first, it was not built on scientific findings in its entirety; second, it would undermine its national interest of continuing to whale. This view on the moratorium was expressed already from the late 1970s onwards when the decision to impose a zero catch quota drew nearer. Once the moratorium was finally adopted, Japan worked tirelessly to overturn it, either by

[2]Sellheim (2016).

lobbying Commission members or by motivating other pro-use countries to join the IWC.

Here, the accusation of vote-buying comes into play since Japan apparently used its foreign aid to lure developing nations into taking a pro-whaling stance in the Commission.[3] From my own observations, this accusation is somewhat difficult to contextualise. While it may have been so in the past, this is certainly not unique. Evidence of secretive meetings does not provide much other evidence than for the holding of strategic meetings amongst like-minded IWC members. This goes both ways and is true for anti-whaling and pro-whaling countries. Moreover, this implies that developing nations are string puppets of the Japanese within the IWC. This is oversimplifying at best, if not offensive. Having attended several of these strategic meetings myself, it is by no means so that developing IWC members unequivocally support Japan. Instead, heated discussions regularly ensue. The common denominator is the sustainable use of resources *vis-à-vis* the no-use agenda. And it is this view that also many developing nations share: they are not necessarily 'pro-whaling', but 'pro-sustainable use'. To merely ascribe them dependency on Japan neglects their capabilities as sovereign nations to stand up for their own interests.

In this context, developing nations within the IWC have also often been criticised for taking a pro-whaling stance yet without having a whaling history themselves. I would like to raise two points in this regard: first, the same can be said about anti-whaling nations. For instance Switzerland, which, at the time of writing chairs the Sub-Committee on Aboriginal Subsistence Whaling, does not have a whaling history—it is a landlocked country—but is stringently anti-commercial whaling. Yet, along with countries that *do* in fact have a whaling history it shares the common denominator of extreme precaution as regards whaling (and sealing, for that matter). Second, a country that is pro-whaling (or pro-sustainable use of resources) does not need to have a whaling history, but may have a whaling future. This is especially true for developing nations which might depend on whale meat as part of their overall food security in the future. It does not come as a surprise, therefore, that it has been developing nations that have put forward the link between the IWC and food security in recent IWC meetings.

Canada is a country that has remained firm concerning its stance on commercial sealing. Canadian commercial sealing is well embedded in the Canadian legal system. Moreover, Canada, along with Norway, has taken steps to defend the commercial seal hunt before the World Trade Organization by trying to overturn the EU Seal Regime. In terms of whaling, Canada has taken the drastic step of leaving the IWC once the moratorium on commercial whaling was adopted in 1982. Although by that time Canada's hunt for large whales had already been non existent anymore, it was fear over the potential extension of the IWC's mandate to include small cetaceans, thus potentially impacting Canada's aboriginal beluga and narwhal hunts.

---

[3]Strand and Tuman (2012).

Canada can be considered a champion of sustainable use of marine resources, including marine mammals. The country is an official observer of NAMMCO, which equally links conservation and sustainable use. Despite Canada leaving the IWC, it has also retained observership in the Commission. It seems, therefore, that Canada follows its own agenda, yet stays involved in the international discourse (and science) on marine mammal utilisation. Opposing the US-led proposal to uplist the polar bear under CITES is but one additional step to substantiate its sustainable use agenda. At the same time, it follows its obligations towards its indigenous population, enshrined in Canadian law. The possibility for Inuit to co-manage polar bear populations as well as to continue their seal and whale hunts substantiates this approach.

The above paragraphs have mostly referred to efforts of states to either allow or restrict hunting operations. However, what all states have in common, be it within the International Whaling Commission, as parties to any sealing agreement, or as parties to the polar bear regime, is the interest in the conservation of marine mammals. This is, after all, the common denominator that all pro- or anti-use states and organisations can agree on. The way by which this happens is, as a result, the subject of disagreement. But as Chap. 4 has demonstrated, species-specific regimes are not the only way in which cooperation (or disagreement) is possible. The OSPAR Convention with its concrete measures to protect the marine environment which I discussed above, or the Convention on Biological Diversity, which I omitted from the discussion, serve as important avenues through which the larger ecosystem-based approach can be taken. Since these regimes are mostly consensus-driven and therefore in the end acceptable to all parties involved, they are significantly less controversial and significantly less infused by questions regarding marine mammal utilisation. Instead, they are regimes which consider the protection of nature as a whole rather than the protection of certain species. Countries like Japan, for instance in the context of the CBD, does not present itself as a pro-whaling nation, but rather as a nation concerned with the conservation of the environment.

8.4 Aboriginal Exemptions

While the way international marine mammal law is shaped, implemented and adjudicated is to a large degree dependent on normative approaches to marine mammal utilisation and associated state behaviour, a rather widely accepted notion is that of aboriginal exemptions. This is to say that many regimes—the ICRW, the EU Seal Regime, the CMS, to name a few—that are still in force include provisions that pay regard to the rights and livelihoods of indigenous (and local) populations. This means that protective measures, predominantly in the form of hunting restrictions, do not, or to a limited degree, apply to indigenous populations.

As I explained in Sect. 1.4, this logic corresponds to the principle of common but differentiated responsibilities. This means that the hunt of aboriginal populations of marine mammals is not considered to be detrimental to the conservation status of a species. Whales, seals and polar bears, in other words, can be hunted on a small scale by indigenous peoples. A common notion that goes hand in hand with this exemption is that of 'subsistence'. Therefore, it is not merely to be a small-scale hunt (which could also be driven by economic considerations), but it is to be motivated by immediate consumption without a market orientation.

In the case of whaling, this approach is somewhat comprehensible as, for instance, Alaskan or Makah whaling takes a small number of whales in the region, makes immediate use of the whale, provides meat and other produce for the entire region and does not have an international—that is to say global—dimension. This stands in stark contrast to the commercial whale hunt, which takes place either in the EEZs or in the high seas, which is conducted by large factory ships, which requires immense technological and financial efforts, which takes larger numbers of whales, and which also holds a global dimension. Irrespective of the question of whether the whaling moratorium is scientifically justifiable or not, from a conservation perspective, aboriginal whaling has a significantly lesser environmental impact than commercial whaling. The same accounted, of course, also to the Bering Sea Fur Seal Regime, the exemption in which was inserted with the same concerns in mind. Although my own interviews with aboriginal representatives at IWC meetings also showed some dissatisfaction with the ASW-principle (the argument being that there should not be a quota for aboriginal whaling in the first place), in general, ASW representatives were rather happy with the fact that despite the global opposition to whaling they are still able to continue with it.

When we turn to the trade in seal products and the EU Seal Regime, however, the picture changes. While meant to spare indigenous populations from any negative effects of a trade barrier, the interlinked market chains have also contributed to negative effects of this ban. I argue that this stems from a rather outdated idea of subsistence sealing. First of all, if sealing in the Arctic or elsewhere were purely subsistence-based, which would mean a purely community-based sphere of exchange, the insertion of any aboriginal exemption into an international trade regime would not be necessary. By inserting it, policy-makers must have been aware of the fact that there is a clear international dimension to subsistence sealing. Problematic in the context of the EU Seal Regime is that the Inuit did not feel adequately consulted by EU policy-makers. Although Inuit organisations vehemently opposed the idea of a seal trade ban, it was put in place anyway. As a consequence, the European Union has been consistently considered as an adversary to Inuit wellbeing since the effects of the 1983 as well as 2009 bans were directly linked to increased suicide rates, loss of culture and identity and associated declining wellbeing. Not surprisingly, counter-campaigns and slogans were developed, such as 'Phoque [the French term for seal] the EU' or 'Save the Veal.' Also the '#sealfie' campaign—a display of seal products in both Inuit and other seal-using cultures—is a direct result of the EU's Seal Regime.

In the case of polar bears, the polar bear regime has from the outset been designed to take into account the interests of polar bear hunters. Both the Polar Bear Agreement as well as the individual management agreements conceptually make extensive use of the concept of aboriginal exemptions. Contrary to other regimes, however, it is not 'exemptions' per se, but rather the incorporation of aboriginal, user interests.

Naturally, in case of the ACPB, the nation state is the sovereign rights holder. After all, the agreement was concluded in the early 1970s. But the conclusion of the other agreements that directly refer to the ACPB shows that its interpretation is being done in a way that puts great emphasis on aboriginal resource rights.

8.5 Non-governmental Influence

In contemporary international marine mammal law, the influence of non-governmental organisations (NGOs) cannot be neglected. With the rise of the environmental movement in the 1960s, non-governmental actors have become part and parcel of the international legal landscape for marine mammals. While they are not decision-makers themselves since the realm of international law is that of nation states, they are nevertheless important actors that have substantially shaped international marine mammal law and international conservation law in general. For example, the 1972 Stockholm conference saw the first attendance of NGOs on a large scale: 170 organisations attended. 40 years later, at the Rio +20 Summit, almost 10,000 NGOs attended the meeting.

NGO-influence has been most prominent in the case of the regimes dealing with international whaling, European Union sealing and trade in wildlife (CITES). In general terms, two overall strategies have been applied by NGOs to influence environmental decision-making: first, through observership and participation in fora of international diplomacy, NGOs have the possibility for direct influence on decision-makers and the generation and transmission of information that might influence state behaviour. The next step includes becoming part of country delegations or other bodies of decision-making. This strategy is referred to as 'inside advocacy'. A second strategy is that of influencing public opinion and discourse on certain issues. Newspaper articles, social media presence or similar means are thus tools to steer public opinion into a certain direction, which would in turn increase pressure on domestic and international decision-makers. This strategy is referred to as 'outside advocacy.'[4]

In the case of sealing, one such example of direct participation in the decision-making process beyond mere influence is that of Brian Davies. Brian Davies is the founder of *IFAW*, an organisation which was formed to work against the commercial seal hunt. While mainly focusing on outside advocacy, over the course of time, Davis had become the principal advisor on seal issues for the Council of the European Communities prior to the adoption of the 1983 Seal Pups Directive. This means that *IFAW* was no longer just in the position to *influence* decision-makers, but had taken up a position to directly *consult* in the decision-making process. Prior to the adoption of the 2009 Seal Regime, *IFAW* and other organisations predominantly focused on outside advocacy and the influencing of decision-makers of the EU. Large demonstrations were staged in Brussels and Strasbourg, stuffed toy

[4]Betzold (2013).

seals were handed to Parliamentarians and other items were distributed to provide for a clear roadmap of how to vote. The overwhelming result of 550 Parliamentarians voting in favour, *vis-à-vis* merely 49 against and 41 abstaining, of the ban on trade in seal products is arguably a direct result of the strategies employed by NGOs. In fact, the support for the ban is one of the highest approval rates in the history of the EU.

Concerning whaling, NGO presence is a common characteristic of the IWC. First of all, over the course of the 1970s, the international, and primarily Western, view on commercial whaling had radically changed. While conservation concerns and associated reductions of whaling activities were one reason, public opinion on whaling due to the generation of the 'superwhale' (see Chap. 1) had led the public to oppose whaling for principal reasons. Interestingly, the capturing of whales and dolphins to be put in aquariums, nowadays an activity that does not find much support, also contributed to this overall change in discourse. With the overall growing of environmental awareness, 'the whale' furthermore changed from the gigantic Leviathan to the fragile species: the giant whale no longer opposed small boats. Instead, the giant whaling vessel opposed small whales. NGOs were crucially involved in advancing this narrative.

Over the course of the 1970s, anti-whaling NGOs became well-funded and well-prepared partakers in the discourse on whaling—both in Western societies as well as within the IWC. The first direct result of this influence was the establishment of the Indian Ocean Whale Sanctuary in 1979, at that time a rather revolutionary approach to whale conservation. In the times prior to the adoption of the moratorium, the IWC saw a significant increase in membership of states without a direct whaling history. Their pay-offs were not significant, but they joined in order to 'fit in'. This shows how influential the NGO-driven discourse on whales and whaling had become. For some countries, NGO members were even part of country delegations. For instance, an *IFAW* activist was member of the German delegation, exerting great influence on German voting behaviour and paving the way for Germany's future anti-whaling stance. In fact, it is the same activist who was crucially responsible for the listing of the minke whale under CITES Appendix I.[5] Also my own observations at the IWC have shown how NGO members were present as NGO delegates on one meeting, but part of country delegations in the next: at IWC67 in 2018 a member of an animal rights NGO was a member of the UK delegation while having been an NGO member at IWC66 in 2016.

In rather recent times, the issue of Japanese whaling has been brought to international attention through televised programmes such as the Oscar-winning film *The Cove* or the action-laden adventure series *Whale Wars*. The former deals with the dolphin drive in Taiji and is the story of US campaigners showcasing the cruelty and supposed secrecy of the drive. After the release of the film in 2008, life in Taiji had changed dramatically, causing an influx of campaigners to stop the drive. Legally, Japan introduced new regulations concerning animal welfare. While in the direct wake of the film hundreds of people annually came to Taiji to protest the hunt, at the

[5]Epstein (2008), p. 159.

time of my own fieldwork during the hunting season in the fall of 2017, merely three campaigners were present. The momentum of the film has passed, it seems. Still, the hunt continues unabated.

Whale Wars, on the other hand, had a more international impact. Telling the story of the *Sea Shepherd Conservation Society* (SSCS) in the Southern Ocean, opposing Japanese scientific whaling, it is inherently biased since it presents merely SSCS' view on the matter. The series stretched over seven seasons from 2008 to 2015 and was aired on Animal Planet, advancing the claim of illegality of Japanese whaling in the Southern Ocean. Both *The Cove* and *Whale Wars* are highly influential televised elements in the anti-whaling outside-advocacy-strategy. While I have yet to come across strategic assessments on their influence, it remains without a doubt that they have left their mark on the public, particularly in Western societies. Thereby, the pressure on both Western governments as well as on the Japanese government to stop all whaling inevitably increased.

The case of the polar bear is somewhat different. NGOs, at least to my knowledge, have not influenced the polar bear regime in the same degree as they have done in the case of whaling and sealing. This may be due to the fact that most NGOs that are opposed to marine mammal hunting primarily do so when the hunt is conducted for commercial reasons. Since polar bears are inherently linked to aboriginal cultures, opposition thereto is rather marginal. This being said, organisations such as *IFAW* or the *Humane Society International* oppose trophy hunting of polar bears. In a joint report, the organisations put forward proof for the lack of tradition and lack of economic importance of polar bear trophy hunting in Canada. In the report it is argued that trophy hunting of polar bears does not alleviate larger problems of Inuit communities and that it is not a mainstay of the Inuit economy.[6] In other words, the underlying narrative appears to support the claim that polar bear trophy hunting is not justifiable from a socio-economic perspective. Without the need to delve deeper into this topic, it suffices to say that the polar bear regime has remained rather unaffected by these efforts.

Given the significant influence of non-governmental organisations on the legal landscape of marine mammal law, it is not surprising that alternative measures have arisen. One of these is the North Atlantic Marine Mammal Commission (NAMMCO) (see Sect. 4.4.2). In our joint article, Marzia Scopelliti and I argue that the mere existence of NAMMCO may point to the fact that some Arctic governments have taken the reigns into their own hands and have formed their own organisation. This happened especially in light of the dysfunction of the IWC and the strong NGO influence in it. As we have shown, although NAMMCO does allow for observers (both governmental and non-governmental) to participate in most meetings, currently no anti-use NGO has officially applied for observer status.[7] Why this is the case, however, I cannot determine. One reason might be that NAMMCO members must uniformly agree on the observer status. It is highly

[6]Waters et al. (2009).

[7]Scopelliti and Sellheim (2019), p. 101.

doubtful that a no-use NGO would be accepted. Another reason could be that by applying for observership, anti-use NGOs would provide NAMMCO with a degree of legitimacy. By ignoring the organisation, NGOs demonstrate their disapproval. Whether the establishment of NAMMCO is an isolated case of rebutting anti-use dominance in the organisations dealing with marine mammals remains to be seen. However, with Japan's withdrawal from the IWC, a new organisation dealing with whales or marine mammals might be a possibility.

8.6 Other Characteristics

As all preceding chapters have demonstrated, international marine mammal law is a scattered body of law—if it can even be considered a separate body of international law. What has become clear, I hope, is that there are different regimes that treat marine mammals under different pretexts. On the one hand, whale protection and utilisation could be globally treated through the International Whaling Commission, depending on the number of contracting parties. Since it is open to a global membership it remains to be seen how its membership will develop in the future. The issue is fundamentally different in the case of seals. Here, it is, and was, rather regional regimes that constitute regulatory efforts.

The question of regulating whales and whaling has first and foremost been a question of regulating the global commons since the commercial whale hunt has predominantly taken place in the high seas. Sealing, on the other hand, has in most cases boiled down to questions relating to the sovereign rights of nation states within the respective maritime zones. The international regulatory regimes relating to small cetaceans resemble sealing regimes in this regard: they are mostly of a regional character while questions of sovereign rights feed into them.

While we have seen that international marine mammal law is in many instances marked by confrontational attitudes or at least by diametrically opposed stances on marine mammal protection, we should not forget that in the past it has also served as a means for deeper cooperation. The Finnish-Russian sealing regimes (see Sect. 6.3.2) or the conclusion of the Polar Bear Agreement show that despite potential conflict, they served as avenues that strengthened the ties between potential adversaries. Of course, it is not these regimes alone that have deepened cooperation, but at least they have opened certain back channels through which additional diplomatic efforts were possible.

All in all, I would argue that international marine mammal law is deeply reactive. Most regimes were concluded when the danger towards a species was already prevalent. These reactive regimes can be subdivided into two distinct strains of regime set-up. On the one hand, more conservative regimes are limited to specific species and do not necessarily take the larger ecosystem-based approach to conservation. Most instruments of the whaling and sealing regimes, the latter including the EU Seal Regime, fall under this category. On the other hand, there are regimes that either take an explicit ecosystem approach, for instance OSPAR, or that are somewhat hybrid in nature. The hybrid regimes can mostly be found in regard to small cetaceans. While they are concluded based on the protection of specific species, they have nevertheless arisen out of regimes that consider species within wider ecosystem considerations. ASCOBANS and ACCOBAMS, that have arisen as Agreements out of the CMS fall into this category.

The Polar Bear Agreement falls in between this distinction, however. First of all, it was concluded in a rather precautionary manner. This means that the immediate threat to polar bears was not apparent at the time of conclusion. As we have seen in Chap. 7, although scientific consensus between the parties did not exist, the agreement was adopted anyway. Through concerted efforts to enable its conclusion, the polar bear agreement is somewhat progressive, also supported by the fact that despite its species-based approach, it nevertheless pays due regard to the ecosystem in which the polar bear occurs.

Over time, also the older, still existent regimes such as the whaling regime or the regime on migratory species has adapted to wider ecosystem-based considerations. In case of the IWC, for example, issues such as noise pollution, climate change or ship strikes now play an increasing role in the work of the commission. Regimes such as the EU Seal Regime, on the other hand, from the outset has merely considered trade in seal products as a source for potential species decline—leaving aside the fact that the EU as changed its course on the *raison d'être* of the regime. Given the regime's set-up it is not possible to include other threats to the species.

In the Antarctic, the sealing regime appears to be more of a complementary, reassuring regime. In a way it resembles the Polar Bear Agreement as it was created when neither seal species were in drastic decline nor an immediate seal hunt was ongoing. It therefore is a precautionary regime that complements the special role of marine mammal regulation within the Antarctic Treaty System.

8.7 Summary and Conclusion

International marine mammal law is a complex body of laws and regulations that has made use of specific characteristics, some of which I have tried to flesh out in this chapter. A crucial element of all regimes is its geographic limitation. Although the whaling convention is global, its original set up made extensive use of geographic boundaries by subdividing whale species into different stocks and allocating certain quotas to these. Also other seemingly global regimes, such as the CMS, are regionally limited due to its different Agreements of a regional scope.

In order to prevent aboriginal populations from being adversely affected by regulatory provisions, aboriginal exemptions have been created. In those areas in which indigenous peoples live, hunting and management as well as the EU Seal Regime have legally considered them to spare them from any adverse effects. I have argued that this corresponds to the principle of common but differentiated responsibilities—a key principle of international environmental law. Whether the concept works in practice, particularly taking into account contemporary socio-economic in indigenous communities, is subject to debate.

What is clear, however, is that most contemporary regimes significantly limit or even oppose commercial pursuit of marine mammals. This stems largely from the fact that non-governmental organisations have successfully applied inside and outside advocacy to influence decision-makers themselves and the public's discourse on marine mammal and marine mammal hunting. Ultimately, of course, it is the nation states that have changed their attitude towards the commercial hunt of marine mammals over the course of the last few decades.

In conclusion, we have seen that international marine mammal law is inherently diverse. It consists of a myriad of international past and present regimes with different foci and institutional settings. To further understand the way marine mammals are internationally protected, their hunt is internationally regulated and to find more commonalities and differences between the different regimes, more research is certainly necessary.

Questions and Research Tasks

- Discuss whether the concept of 'international marine mammal law' is adequate.
- Think of more common elements as well as differences between the legal regimes presented in this book.
- Are there specific trends that you can determine when considering the different regimes?
- What are the interplays between the regimes that constitute international marine mammal law and other spheres of international law?

References

Betzold C (2013) Business insiders and environmental outsiders? Advocacy strategies in international climate change negotiations. Int Group Adv 2–3:302–322

Epstein C (2008) The power of words in international relations. Birth of an anti-whaling discourse. The MIT Press, Cambridge

Scopelliti M, Sellheim N (2019) Sustaining a conservationist agenda? NGO influence on Arctic sealing, whaling and hydrocarbon regimes. In: Shibata A, Zou L, Sellheim N, Scopelliti M (eds) Emerging legal orders in the Arctic. The role of non-Arctic actors. Routledge, Abingdon, pp 91–106

Sellheim N (2016) The legal question of morality: seal hunting and the European moral standard. Soc Legal Stud 25(2):141–161

Strand JR, Tuman JP (2012) Foreign aid and voting behavior in an international organization: the case of Japan and the international whaling commission. FPA 8(4):409–430

Waters M, Rose N, Todd P (2009) The economics of polar bear trophy hunting in Canada. IFAW, Yarmouth Port. https://www.zeit.de/wissen/umwelt/2013-03/trophy-hunting-ifaw.pdf. Accessed 11 June 2019

WTO (1996) European Communities — Measures Concerning Meat and Meat Products (Hormones), DS26

The Future of International Marine Mammal Law

9

9.1 Introduction

What we have noted throughout this book are two essential facets that characterise international marine mammal law: first, it is extremely scattered; second, its controversial, adversarial nature. With Japan leaving the International Whaling Commission, a new era of whaling regulation has begun. Whether this will send ripple effects through the legal landscape of marine mammal conservation efforts remains to be seen. However, what it does show is that Japan—and possibly others—do not consider the IWC a legitimate organisation anymore to represent the interests of all its members. But where does this lead? In this chapter, I try to sketch a few scenarios which might be a possibility in light of the current developments in international marine mammal law. I am drawing from the existing literature and from the foregoing chapters. My own view on different scenarios based on the above will inevitably flow into this chapter. However, you may disagree with my assessment and I strongly encourage you to develop and communicate your own ideas as regards the future of international marine mammal law.

9.2 Scenario 1: Same Old, Same Old

In this scenario, not much will change. Although Japan has left the International Whaling Commission, the IWC will persist in its current form, may lose or gain some members and will proceed on the trajectory towards a more conservationist approach to whale management. Japan will remain an observer to the Commission and will cooperate with its Scientific Committee, but since the IWC will retain, or even strengthen, its no-use agenda of whales, Japan will not rejoin again. This would also mean that Japan conducts commercial whaling within its own EEZ. Until 2020, Japan will also conduct research whaling in the North Pacific, when, as the state of affairs at the time of writing indicates, all Japanese research whaling is suspended. Both types of whaling can hardly be legally challenged. On the one hand, since

N. Sellheim, *International Marine Mammal Law*,
https://doi.org/10.1007/978-3-030-35268-4_9

Japan is still cooperating with the IWC, especially with its Scientific Committee, it is not in breach of the cooperation requirement set out in the UNCLOS. One possibility that might arise is that a notification of Japan occurs under the Pelly Amendment under US law. This would mean that economic sanctions might be imposed by the US in order to force Japan to abandon its whaling activities (see Sect. 5.3.7). Under the Trump Administration this is highly unlikely since also other administrations that have followed stricter environmental principles have thus far abstained from such notification.

With Japan's withdrawal from the IWC, an outspoken pro-whaling country has left the Commission. With it, also one of the most solvent contributors to the Commission has left, too, particularly with regard to Japan's contributions to the Scientific Committee. In how far this will affect the work of the IWC is difficult to foresee. One way or another, however, the IWC will have to scale down its work in some way. Where within the Commission this scaling down will occur would be purely speculative at this point. What is more interesting is the question what will happen to the so-called 'sustainable use group' in the Commission. After all, quite a large number of countries—around 25—including Norway, Iceland, the Russian Federation, St Vincent & the Grenadines, Antigua & Barbuda, Liberia or Senegal, have now lost a powerful and financially capable ally. Particularly for Caribbean and African states, this might lead to the question in how far membership without Japan is still a feasible and economic option. Therefore, it would not come as a surprise if some of these countries would leave the Commission as well, particularly since without Japan, the most powerful voice for sustainable use has been silenced. In this way, the question of dysfunctionality would somewhat resolve itself: from a conservation perspective, the IWC would be perfectly functional without the 'dissenting' voices of sustainable use countries. Therefore, if more sustainable use countries were to leave the Commission, the IWC would become a rather well-functioning conservation organisation, irrespective of whether it set out to be such an organisation from the very beginning or not. In the longer run, however, his would hurt the IWC since potentially more and more whaling would occur outside of its auspices, inevitably leading to decreasing effectiveness and an overall decline in international legitimacy.

The developments in the IWC would not affect other regimes. The CMS and its Agreements, the Antarctic sealing convention, the EU Seal Regime, the Polar Bear Regime and others would remain in place and would continue to exist unabatedly. If the IWC were to extend its mandate to include small cetaceans, however, it would be particularly the CMS and its Agreements which would further deepen its cooperative ties with the IWC. Also potential conflict with NAMMCO might arise as to the increasingly overlapping mandates of both organisations. However, since also conservation-oriented members of the IWC have not agreed to an extension of its mandate, this is rather unlikely, albeit not impossible.

The only regime that might be more directly affected by the developments in the IWC is CITES. Also in the past, decisions in the IWC affected CITES and vice versa. In fact, some of the actors that are present at the IWC and at CITES meetings are the same. Instead of the IWC, it might be CITES which would become the battleground between pro-sustainable use and anti-sustainable use countries. While, as we have seen, Japan, Iceland and others have lodged reservations to the listing of whale species on CITES Appendix I and trade in these products is therefore legal between these countries (Sect. 4.2.1), this might not prevent other countries and NGOs to increase their campaigns to label, for instance, Japanese and Icelandic whale meat trade as illegal. Whether, in the long run, this would ultimately lead to a similar situation of deadlock in CITES as in the IWC remains to be seen.

9.3 Scenario 2: The Moratorium Is Lifted, the Revised Management Scheme Put in Place

In the second scenario we look at the possibility of the Revised Management Scheme being put in place. This would effectively mean that the moratorium on commercial whaling is lifted. While under current circumstances this scenario is highly unlikely, it is still worth considering. The reason is that in recent years, the IWC has seen an increasing discourse on food security. While at the time of writing no official resolution has been adopted, in light of the UN's Sustainable Development Goals that see Zero Hunger as their second goal,[1] it appears that irrespective of the commercial aspects of whaling, whales may also start to be considered as a legitimate food source by the international community. In essence, this is already occurring on the level of Aboriginal Subsistence Whaling, yet has thus far failed to enter the wider discourse on food security.

If—or even when—the world, due to climate change or other environmental disasters, is going to experience increasing food shortages, it appears almost inevitable that increasing attention would be shifted towards all living resources roaming the oceans. This might particularly affect African and Caribbean nations, especially since decreasing fish stocks in the Atlantic might force governments to consider other food sources, for instance whales. If that might happen, also the debate on whether or not the moratorium on commercial whaling is still justifiable and whether a lift might be reasonable will open up again. It is almost certain that developed states, such as the US, EU member states or Australia and New Zealand would still vehemently oppose such proposal. After all, it is these countries that have thus far opposed the opening of a strategic discourse on food security within the IWC. Notwithstanding, if also these countries were to finally recognise whales as a legitimate food source, attention would in all likelihood be turned to the Revised Management Scheme (RMS) and very careful, precautionary modes of whale management. With the implementation of the RMS, the zero catch quota of the Schedule would be lifted and limited, strictly science-based commercial whaling would resume. Once again, we must consider whether the term 'commercial' would be adequate in this context. While due to their size and high seas habitats whales naturally require financial efforts that can only be carried by large companies, it would be fair to assume that a resumption of whaling for food would not lead to new

[1]United Nations (Undated).

'Whaling Olympics'. The basis for the assumption lies in yet another assumption, namely that the IWC would, first, set stringent quotas and, second, would allow whaling merely for specific domestic markets. This means that countries that do not have the required skill or economic capacities would enter into bilateral contracts with economically more powerful whaling nations so that they can provide them with a certain amount of whale meat.

If this scenario were to take place, it would appear almost inevitable that Japan would rejoin the Commission since it would be interested in jointly managing global commons. Although Japan has always followed its own national interest in the Commission, it has also been a champion of cooperation as regards whale management. There would be no reason for Japan not to join the IWC anymore once limited commercial whaling would be possible.

If the moratorium were to be lifted, this would probably send ripple effects throughout the entire landscape of international marine mammal law and the law of the sea. The reason is that this would mean a *de facto* recognition of marine mammals as a legitimate food source by the international community. Depending on the status of food security, also other regions might start to consider seals or polar bears as food. Therefore, quite possibly regimes such as the EU Seal Regime or CITES would be subjected to increasing pressure from the public. However, while these regimes are rather tightly regulated and enforceable, the situation is different concerning the regimes for small cetaceans. It is possible that in particular ACCOBAMS and ASCOBANS might suffer from this development since there is potential for increasingly unsustainable use of cetaceans. After all, these regimes are, first, much lesser known than the IWC, CITES and EU Seal Regime and, second, there are no ways to enforce them. This is especially so if range states, but not member states, to these regimes conduct larger scale hunts for small cetaceans.

In order to counteract the potential for unsustainable hunts it would therefore be up to the national governments to enforce rules for sustainable use. In order to push this approach, new non-binding, 'soft law' agreements could be taken that combine conservation and sustainable use, but which leave it up to the nation states to enforce them. The benefit of soft law agreements over legally binding ones is that they do not carry obligations, but set out normative standards and codes of practice. This would provide for new and innovative ways to integrate different views on marine mammal conservation and utilisation to be applied. Presumably, this would not occur on a global scale, but rather at regional levels.

However, in order to regulate marine mammal conservation even in light of a lifting of the moratorium, it is also possible that different states enter into 'smaller' agreements for their management. This might occur on a species basis or a regional basis. The blueprint for such agreements might stem from the already existing Regional Fisheries Management Organisations (RFMOs). At the same time it would not be surprising if the IWC were to see an increase in membership, particularly by those states potentially wishing to engage in whaling to provide food. Indonesia and Canada might be two of such countries. While the former has never been a member of the IWC, but still conducts whaling for great whales, the latter left the IWC due to the impending moratorium. With the moratorium lifted, Canada might be motivated again to cooperate with other states on the sustainable use of whales.

9.4 Scenario 3: A New Organisation Is Formed

A third scenario involved the formation of a new organisation. The driving force behind this endeavour would be Japan, which would be interested in forming either a regional or even a global organisation for the conservation and sustainable use of whales (and other species). If the organisation were to be global, problematic in this regard might be the overlapping competences of the IWC and the new organisation. Therefore, it appears more reasonable to assume that if Japan were to motivate the formation of a new organisation, this would be driven by the principle of sustainable use of marine (and possibly other) resources. This could include the possibility for all nations to be able to join. But since, as we have seen in Sect. 5.4, particularly many nations have established *de facto* protected geographical and legal areas for whales and other marine mammals, this new organisation might be open to nations that are explicitly interested in the conservation *and* utilisation of marine mammals. This might include Asian states such as China and South Korea, along with Micronesian and Melanesian, Caribbean and African governments.

Since the new organisation would be global, it would be open for all states to join. To further bolster the spirit of cooperation as stipulated in the UNCLOS, the organisation would be open for observers—both governmental and non-governmental. However, as is the case with NAMMCO, observers, particularly non-governmental organisations, would have to adhere to a strict code of conduct that ensures the smooth functioning of the organisation's meetings. Decisions of the organisation and within its potential sub-committees would be taken by consensus in order to ensure that all member states are content with the decisions that are made. Whether or not all member states would actively engage in whaling is, of course, doubtful. However, the organisation might be relevant for the defence of the principle of sustainable use of resources.

Yet, in order to enable this, all members of the organisation would have to lodge formal reservations against the listing of marine mammal species under CITES. As we have seen in Sect. 4.2.1, Palau has already done so and would therefore be able to enter into a trade agreement with Japan on certain whale species.

9.5 Scenario 4: The Formation of an International Marine Mammal Commission

In this scenario I briefly summarise and present the idea that Cameron Jefferies put forward in his excellent treatise *Marine Mammal Conservation and the Law of the Sea*.[2] In his book, Jefferies puts great effort into the idea of an International Marine Mammal Commission (IMMC). As the legal basis for this idea serve articles 65 and 120 of the UNCLOS since they require states to cooperate on the management and conservation of marine mammals, both in EEZs as well as in the high seas (see Sect. 4.2.3). The underlying idea, of course, is that the current legal framework for marine mammal conservation is not sufficient and that it is in need of reform. The justification for this assessment lies in three characteristics: the framework is too fragmented; it is insufficient in terms of species coverage and threat inclusion; and it is outdated since modern principles of conservation are not included.

Yet before we enter into the discussion, let me briefly outline why I think, it is unlikely that the world will see an international treaty on seals and sealing. Throughout the book I hope I have demonstrated that there is a key difference between whaling and sealing regimes: contrary to whaling, sealing occurs in areas closer to land (if not even on land) and is therefore subject to national regulation. In the past, regional regimes existed, but in those areas in which sealing is still actively being conducted, this occurs in the sovereign territories of nation states. In the Antarctic, sealing is regulated by the Convention on the Conservation of Antarctic Seals. Therefore, there is no need to further advance the idea of a sealing treaty for the Antarctic. Moreover, trade in seal products has already declined significantly due to the collapse of international markets. Economically, the seal hunt does therefore not play such an important role anymore which would warrant the conclusion of a specific sealing treaty.

Enters the proposal put forward by Cameron Jefferies. In light of the current situation, he proposes the formation of a new treaty, a UN Marine Mammal Agreement. The main purpose of this new regime would be the rational management and long-term conservation of marine mammals worldwide. However, the International Marine Mammal Commission, which would be established under this regime, would have dual competencies: on the one hand, it would have primary power to take decisions over marine mammal conservation and utilisation on the high seas; on the other, it would require the consent of the coastal state to make regulatory decisions over its EEZs.

With this approval, the IMMC would be able to implement management advice from its Scientific Committee. These decisions would all be taken by consensus. While states would be able to express their dissatisfaction with specific decisions, Jefferies' point of departure is that there would not be a formal way of objecting to a decision or to lodge official reservations. A special role that would fall onto the Scientific Committee would be able to propose the establishment of marine protected

[2]Jefferies (2016).

areas for marine mammals both within national jurisdiction and in the high seas. Contrary to the concept of whale sanctuaries, this would include close cooperation with the local population and would not exclude the sustainable use of marine mammals.

As we have seen in Sect. 4.2.3, under the UNCLOS marine mammals take a special place. Some species are listed in Annex I as highly migratory species warranting special protection. Jefferies proposes the removal of these species from the Annex and to place them directly under the competence of the IMMC. With the experiences of the IWC in mind, it appears almost inevitable that within the IMMC disputes on conservation and sustainable use would arise. In order to prevent a stall such as in the IWC, the UN Marine Mammal Agreement would have clearly set out dispute resolution mechanisms that draw from the UNCLOS as well as the UN Fish Stocks Agreement and would include "negotiation, inquiry, mediation, conciliation, arbitration, judicial settlement, resort to [the IMMC or] regional agencies or arrangements, or other peaceful means of their choice."[3]

While these steps appear reasonable and Jefferies further lays out in great detail the benefits of such new agreement, he also concedes the fact that this new agreement would face significant political obstacles. I would certainly agree with this assessment. After all, all members of the IWC have sufficient experience with the lack of cooperative will from the respective other side. Japan, Norway and Iceland might fear that despite dispute resolution measures there will be an eventual stall on cooperation. Therefore, it would be preferable for them to either continue trying to lift the moratorium, to leave the IWC or to start a new organisation. After all, Norway and Iceland both have lodged objections to the moratorium and are therefore entitled to conduct whaling. Also, an amendment to the UNCLOS appears highly unlikely. The decade-long negotiation phase demonstrate how difficult achieving consensus on this agreement was. In order to avoid amendments, parties to UNCLOS consequently rely on implementing agreements rather than having to go through an amendment procedure.

Those states that have put the moratorium in place and have maintained it to this day may be deterred by the fact that a new organisation would take precedent over the whaling convention. This would also *de facto* mean that commercial whaling would be possible again, not only by Japan, Norway and Iceland, but also by other states. Moreover, the IMMC would not only include whales, but all marine mammals. I do not see a reason for a country like Canada to potentially subdue some of its decision-making authority to the Scientific Committee of the IMMC and have its seal hunting and marine mammal policies further scrutinised by the international community. Also regarding polar bears, a new agreement would rather undermine the user-to-user approach of polar bear management than strengthen cooperative behaviour. After all, it would be nation states from other regions of the world that would exert influence on the way polar bears are managed—at least within the EEZ. Although Jefferies envisions an Indigenous and Cultural/Artisanal

[3]Jefferies (2016), p. 243.

Subsistence Take Committee, I doubt that countries would be willing to accept this. A Japanese official expressed to me, for instance, that this would preclude the consideration of whales as a future food source for countries that do not have a whaling tradition or whaling culture. Therefore, whaling considerations should not pay too much attention to cultural issues.

While Jefferies' proposal holds extremely valuable considerations, I would assume that it would correspond more to an 'ideal world' than to the bitter realities of marine mammal management and conservation. After all, reason and science do not prevail in the discourse on marine mammals. We therefore cannot assume that an organisation which would be fully rested in reason and science would emerge in the future.

9.6 Scenario 5: Marine Mammals Increasingly Find Their Way Into Other Regimes

The last scenario I briefly wish to discuss is, given the fragmented nature of contemporary international marine mammal law, the possibility of marine mammal regulation finding its way into other regimes. As the observant reader may have noticed, in Chap. 4 I omitted the Convention on Biological Diversity (CBD). This was deliberate. I chose to do so since within the CBD, marine mammals play a rather non-existing role. Much more attention is paid to overall strategies for conservation and sustainable use of biodiversity instead of to specific species—after all the CBD is not species-based. Of course, marine mammals are part of this approach, yet contrary to other regimes, the CBD is significantly less confrontational, consensus-based and a regime that sets international standards but does not make management decisions.

Given the history of cooperative spirit in the CBD, issues related to marine protected areas for marine mammals or the inclusion of local communities for marine mammal management, particularly whales, might occur in the future, in case the IWC loses its authority. In this case it is also possible that the UN Fish Stocks Agreement might make increasing reference to marine mammals and that the responsible regional fisheries organisations might also extend to marine mammals. Also the ongoing conference on an agreement on biodiversity beyond national jurisdiction, a process under the UNCLOS, might make marine mammal conservation and utilisation a part of the discussions. Additionally, the climate change regime, shipping and pollution regimes that have thus far not considered marine mammals, might make increasing reference to them and even develop concrete means for their protection. However unlikely, it all depends on the way the IWC develops in the future and whether or not any of the other scenarios that I sketched above (or any scenario that I have not considered) might come true.

9.7 Summary and Conclusion

Whether or not any of the scenarios above will become reality remains of course in the realm of speculation. I chose to include this chapter, however, to point to the fact that currently the world appears to be at a crossroads regarding international marine mammal law. On the one hand, whales, seals and polar bears will remain iconic species that will always retain a special place in international law-making. On the other, the approach to environmental conservation is currently changing and more and more ecosystem-based management is appearing. This means that marine mammal species are no longer singled out, but considered within a wider environmental perception. This perception does include humans as well. The consideration of humans as part of the ecosystem is, of course, not new and has been part and parcel of indigenous worldviews and mythologies. However, in the western world, this link has been coined a 'socio-ecological system' in which both nature and culture depend on one another.[4]

With the presentation of these scenarios I end this book. Ideally, I would hope to see international marine mammal law develop in a way that does indeed see all different interests combined. This would essentially also mean that some degree of whaling, sealing and polar bear hunting and trade is possible. None of this, however, should occur at the expense of the species in question. That is to say, the sustainability of marine mammal populations should be the highest good, rooted in the principle of precaution. Blanket bans on either hunting or trade, however, can be counterproductive as they do not motivate dissenting views to be accommodated.

With the multitude of threats towards marine mammals, I am quite certain that they will not disappear from international attention. Furthermore, marine mammals will certainly not disappear as iconic species. With this book I hope I have managed to shed some light on the misconceptions on marine mammal and international marine mammal law. Yet, no book is ever complete and it is possible that within a short period of time it becomes clear that neither of my scenarios is a realistic one. Time will tell.

Questions and Research Tasks

- In your opinion, what might the future of international marine mammal law look like? Are there other scenarios that you could envision?
- Which of the five scenarios presented here is the most reasonable and which one is not?
- Are there any other ways to overcome differences in perception on marine mammals?

[4]Berkes et al. (2003).

References

Berkes F, Colding J, Folke C (2003) Navigating social–ecological systems: building resilience for complexity and change. Cambridge University Press, Cambridge

Jefferies CSG (2016) Marine mammal conservation and the law of the sea. Oxford University Press, Oxford

United Nations (Undated) About the Sustainable Development Goals. https://www.un.org/sustainabledevelopment/sustainable-development-goals/. Accessed 12 June 2019

International Convention for the Regulation of Whaling, 1946 (as amended by Protocol of 19 November 1956)

The Governments Whose Duly Authorised Representatives Have Subscribed Hereto

Recognizing the interest of the nations of the world in safeguarding for future generations the great natural resources represented by the whale stocks;

Considering that the history of whaling has seen over-fishing of one area after another and of one species of whale after another to such a degree that it is essential to protect all species of whales from further over-fishing;

Recognizing that the whale stocks are susceptible of natural increases if whaling is properly regulated, and that increases in the size of whale stocks will permit increases in the number of whales which may be captured without endangering these natural resources;

Recognizing that it is in the common interest to achieve the optimum level of whale stocks as rapidly as possible without causing widespread economic and nutritional distress;

Recognizing that in the course of achieving these objectives, whaling operations should be confined to those species best able to sustain exploitation in order to give an interval for recovery to certain species of whales now depleted in numbers;

Desiring to establish a system of international regulation for the whale fisheries to ensure proper and effective conservation and development of whale stocks on the basis of the principles embodied in the provisions of the International Agreement for the Regulation of Whaling, signed in London on 8th June, 1937, and the protocols to that Agreement signed in London on 24th June, 1938, and 26th November, 1945; and

Having decided to conclude a convention to provide for the proper conservation of whale stocks and thus make possible the orderly development of the whaling industry;

N. Sellheim, *International Marine Mammal Law*,
https://doi.org/10.1007/978-3-030-35268-4

Have agreed as follows:-

Article I

1. This Convention includes the Schedule attached thereto which forms an integral part thereof. All references to "Convention" shall be understood as including the said Schedule either in its present terms or as amended in accordance with the provisions of Article V.
2. This Convention applies to factory ships, land stations, and whale catchers under the jurisdiction of the Contracting Governments and to all waters in which whaling is prosecuted by such factory ships, land stations, and whale catchers.

Article II

As used in this Convention:-

1. "Factory ship" means a ship in which or on which whales are treated either wholly or in part;
2. "Land station" means a factory on the land at which whales are treated whether wholly or in part;
3. "Whale catcher" means a helicopter, or other aircraft, or a ship, used for the purpose of hunting, taking, killing, towing, holding on to, or scouting for whales;
4. "Contracting Government" means any Government which has deposited an instrument of ratification or has given notice of adherence to this Convention.

Article III

1. The Contracting Governments agree to establish an International Whaling Commission, hereinafter referred to as the Commission, to be composed of one member from each Contracting Government. Each member shall have one vote and may be accompanied by one or more experts and advisers.
2. The Commission shall elect from its own members a Chairman and Vice-Chairman and shall determine its own Rules of Procedure. Decisions of the Commission shall be taken by a simple majority of those members voting except that a three-fourths majority of those members voting shall be required for action in pursuance of Article V. The Rules of Procedure may provide for decisions otherwise than at meetings of the Commission.
3. The Commission may appoint its own Secretary and staff.
4. The Commission may set up, from among its own members and experts or advisers, such committees as it considers desirable to perform such functions as it may authorize.
5. The expenses of each member of the Commission and of his experts and advisers shall be determined by his own Government.

6. Recognizing that specialized agencies related to the United Nations will be concerned with the conservation and development of whale fisheries and the products arising there from and desiring to avoid duplication of functions, the Contracting Governments will consult among themselves within 2 years after the coming into force of this Convention to decide whether the Commission shall be brought within the framework of a specialized agency related to the United Nations.
7. In the meantime the Government of the United Kingdom of Great Britain and Northern Ireland shall arrange, in consultation with the other Contracting Governments, to convene the first meeting of the Commission, and shall initiate the consultation referred to in paragraph 6 above.
8. Subsequent meetings of the Commission shall be convened as the Commission may determine.

Article IV

1. The Commission may either in collaboration with or through independent agencies of the Contracting Governments or other public or private agencies, establishments, or organizations, or independently:
   (a) encourage, recommend, or if necessary, organize studies and investigations relating to whales and whaling;
   (b) collect and analyze statistical information concerning the current condition and trend of the whale stocks and the effects of whaling activities thereon;
   (c) study, appraise, and disseminate information concerning methods of maintaining and increasing the populations of whale stocks.
2. The Commission shall arrange for the publication of reports of its activities, and it may publish independently or in collaboration with the International Bureau for Whaling Statistics at Sandefjord in Norway and other organizations and agencies such reports as it deems appropriate, as well as statistical, scientific, and other pertinent information relating to whales and whaling.

Article V

1. The Commission may amend from time to time the provisions of the Schedule by adopting regulations with respect to the conservation and utilization of whale resources, fixing:
   (a) protected and unprotected species;
   (b) open and closed seasons;
   (c) open and closed waters, including the designation of sanctuary areas;
   (d) size limits for each species;
   (e) time, methods, and intensity of whaling (including the maximum catch of whales to be taken in any one season);

(f) types and specifications of gear and apparatus and appliances which may be used;
(g) methods of measurement;
(h) catch returns and other statistical and biological records; and
(i) methods of inspection.

2. These amendments of the Schedule:
   (a) shall be such as are necessary to carry out the objectives and purposes of this Convention and to provide for the conservation, development, and optimum utilization of the whale resources;
   (b) shall be based on scientific findings;
   (c) shall not involve restrictions on the number or nationality of factory ships or land stations, nor allocate specific quotas to any factory or ship or land station or to any group of factory ships or land stations; and
   (d) shall take into consideration the interests of the consumers of whale products and the whaling industry.
3. Each of such amendments shall become effective with respect to the Contracting Governments 90 days following notification of the amendment by the Commission to each of the Contracting Governments, except that:
   (a) if any Government presents to the Commission objection to any amendment prior to the expiration of this 90-day period, the amendment shall not become effective with respect to any of the Governments for an additional 90 days;
   (b) thereupon, any other Contracting Government may present objection to the amendment at any time prior to the expiration of the additional 90-day period, or before the expiration of 30 days from the date of receipt of the last objection received during such additional 90-day period, whichever date shall be the later; and
   (c) thereafter, the amendment shall become effective with respect to all Contracting Governments which have not presented objection but shall not become effective with respect to any Government which has so objected until such date as the objection is withdrawn. The Commission shall notify each Contracting Government immediately upon receipt of each objection and withdrawal and each Contracting Government shall acknowledge receipt of all notifications of amendments, objections, and withdrawals.
4. No amendments shall become effective before 1st July, 1949.

Article VI

The Commission may from time to time make recommendations to any or all Contracting Governments on any matters which relate to whales or whaling and to the objectives and purposes of this Convention.

Article VII

The Contracting Government shall ensure prompt transmission to the International Bureau for Whaling Statistics at Sandefjord in Norway, or to such other body as the Commission may designate, of notifications and statistical and other information required by this Convention in such form and manner as may be prescribed by the Commission.

Article VIII

1. Notwithstanding anything contained in this Convention any Contracting Government may grant to any of its nationals a special permit authorizing that national to kill, take and treat whales for purposes of scientific research subject to such restrictions as to number and subject to such other conditions as the Contracting Government thinks fit, and the killing, taking, and treating of whales in accordance with the provisions of this Article shall be exempt from the operation of this Convention. Each Contracting Government shall report at once to the Commission all such authorizations which it has granted. Each Contracting Government may at any time revoke any such special permit which it has granted.
2. Any whales taken under these special permits shall so far as practicable be processed and the proceeds shall be dealt with in accordance with directions issued by the Government by which the permit was granted.
3. Each Contracting Government shall transmit to such body as may be designated by the Commission, in so far as practicable, and at intervals of not more than 1 year, scientific information available to that Government with respect to whales and whaling, including the results of research conducted pursuant to paragraph 1 of this Article and to Article IV.
4. Recognizing that continuous collection and analysis of biological data in connection with the operations of factory ships and land stations are indispensable to sound and constructive management of the whale fisheries, the Contracting Governments will take all practicable measures to obtain such data.

Article IX

1. Each Contracting Government shall take appropriate measures to ensure the application of the provisions of this Convention and the punishment of infractions against the said provisions in operations carried out by persons or by vessels under its jurisdiction.
2. No bonus or other remuneration calculated with relation to the results of their work shall be paid to the gunners and crews of whale catchers in respect of any whales the taking of which is forbidden by this Convention.
3. Prosecution for infractions against or contraventions of this Convention shall be instituted by the Government having jurisdiction over the offence.
4. Each Contracting Government shall transmit to the Commission full details of each infraction of the provisions of this Convention by persons or vessels under the jurisdiction of that Government as reported by its inspectors. This information shall include a statement of measures taken for dealing with the infraction and of penalties imposed.

Article X

1. This Convention shall be ratified and the instruments of ratifications shall be deposited with the Government of the United States of America.
2. Any Government which has not signed this Convention may adhere thereto after it enters into force by a notification in writing to the Government of the United States of America.
3. The Government of the United States of America shall inform all other signatory Governments and all adhering Governments of all ratifications deposited and adherences received.
4. This Convention shall, when instruments of ratification have been deposited by at least six signatory Governments, which shall include the Governments of the Netherlands, Norway, the Union of Soviet Socialist Republics, the United Kingdom of Great Britain and Northern Ireland, and the United States of America, enter into force with respect to those Governments and shall enter into force with respect to each Government which subsequently ratifies or adheres on the date of the deposit of its instrument of ratification or the receipt of its notification of adherence.
5. The provisions of the Schedule shall not apply prior to 1st July, 1948. Amendments to the Schedule adopted pursuant to Article V shall not apply prior to 1st July, 1949.

Article XI

Any Contracting Government may withdraw from this Convention on 30th June, of any year by giving notice on or before 1st January, of the same year to the depository Government, which upon receipt of such a notice shall at once communicate it to the other Contracting Governments. Any other Contracting Government may, in like manner, within one month of the receipt of a copy of such a notice from the depository Government give notice of withdrawal, so that the Convention shall cease to be in force on 30th June, of the same year with respect to the Government giving such notice of withdrawal.

The Convention shall bear the date on which it is opened for signature and shall remain open for signature for a period of 14 days thereafter.

In witness whereof the undersigned, being duly authorized, have signed this Convention.

Done in Washington this second day of December, 1946, in the English language, the original of which shall be deposited in the archives of the Government of the United States of America. The Government of the United States of America shall transmit certified copies thereof to all the other signatory and adhering Governments.

SIGNATORIES:

FOR ARGENTINA:	Oscar Ivanissevich, José Manuel Moneta, Guillermo Brown, Pedro H. Bruno Videla
FOR AUSTRALIA:	Francis F. Anderson

FOR BRAZIL:	Paulo Fróes da Cruz
FOR CANADA:	H.H. Wrong, H.A. Scott
FOR CHILE:	Augustín R. Edwards
FOR DENMARK:	Peter Friedrich Erichsen
FOR FRANCE:	Francis Lacoste
FOR THE NETHERLANDS:	Guy Richardson Powles
FOR NEW ZEALAND:	Birger Bergersen
FOR PERU:	Carlos Rotalde
FOR THE UNION OF SOVIET SOCIALIST REPUBLICS:	Alexander S. Bogdanov, Eugine I. Nikishin
FOR THE UNITED KINGDOM OF GREAT BRITAIN AND NORTHERN IRELAND:	A.T.A. Dobson, J. Thomson
FOR THE UNITED STATES OF AMERICA:	Remington Kellogg, Ira N. Gabrielson, William E.S. Flory
FOR THE UNION OF SOUTH AFRICA:	H.T. Andrews

Agreement for the Conservation of Polar Bears, 1973

The Governments of Canada, Denmark, Norway, the Union of Soviet Socialist Republics and the United States of America

Recognizing the special responsibilities and special interests of the States of the Arctic Region in relation to the protection of the fauna and flora of the Arctic Region;

Recognizing that the polar bear is a significant resource of the Arctic Region which requires additional protection;

Having decided that such protection should be achieved through co-ordinated national measures taken by the States of the Arctic Region;

Desiring to take immediate action to bring further conservation and management measures into effect;

Having agreed as follows:

Article I

1. The taking of polar bears shall be prohibited except as provided in Article III.
2. For the purposes of this Agreement, the term "taking" includes hunting, killing and capturing.

Article II

Each Contracting Party shall take appropriate action to protect the ecosystems of which polar bears are a part, with special attention to habitat components such as denning and feeding sites and migration patterns, and shall manage polar bear populations in accordance with sound conservation practices based on the best available scientific data.

Article III

1. Subject to the provisions of Articles II and IV any Contracting Party may allow the taking of polar bears when such taking is carried out:
   (a) for bona fide scientific purposes; or

N. Sellheim, *International Marine Mammal Law*,
https://doi.org/10.1007/978-3-030-35268-4

(b) by that Party for conservation purposes; or
(c) to prevent serious disturbance of the management of other living resources, subject to forfeiture to that Party of the skins and other items of value resulting from such taking; or
(d) by local people using traditional methods in the exercise of their traditional rights and in accordance with the laws of that Party; or
(e) wherever polar bears have or might have been subject to taking by traditional means by its nationals.

2. The skins and other items of value resulting from taking under sub-paragraph (b) and (c) of paragraph 1 of this Article shall not be available for commercial purposes.

Article IV
The use of aircraft and large motorized vessels for the purpose of taking polar bears shall be prohibited, except where the application of such prohibition would be inconsistent with domestic laws.

Article V
A Contracting Party shall prohibit the exportation from, the importation and delivery into, and traffic within, its territory of polar bears or any part or product thereof taken in violation of this Agreement.

Article VI

1. Each Contracting Party shall enact and enforce such legislation and other measures as may be necessary for the purpose of giving effect to this Agreement.
2. Nothing in this Agreement shall prevent a Contracting Party from maintaining or amending existing legislation or other measures or establishing new measures on the taking of polar bears so as to provide more stringent controls than those required under the provisions of this Agreement.

Article VII
The Contracting Parties shall conduct national research programmes on polar bears, particularly research relating to the conservation and management of the species. They shall as appropriate co-ordinate such research with research carried out by other Parties, consult with other Parties on the management of migrating polar bear populations, and exchange information on research and management programmes, research results and data on bears taken.

Article VIII
Each Contracting Party shall take action as appropriate to promote compliance with the provisions of this Agreement by nationals of States not party to this Agreement.

Article IX

The Contracting Parties shall continue to consult with one another with the object of giving further protection to polar bears.

Article X

1. This Agreement shall be open for signature at Oslo by the Governments of Canada, Denmark, Norway, the Union of Soviet Socialist Republics and the United States of America until 31st March 1974.
2. This Agreement shall be subject to ratification or approval by the signatory Governments. Instruments of ratification or approval shall be deposited with the Government of Norway as soon as possible.
3. This Agreement shall be open for accession by the Governments referred to in paragraph I of this Article. Instruments of accession shall be deposited with the Depositary Government.
4. This Agreement shall enter into force 90 days after the deposit of the third instrument of ratification, approval or accession. Thereafter, it shall enter into force for a signatory or acceding Government on the date of deposit of its instrument of ratification. approval or accession.
5. This Agreement shall remain in force initially for a period of 5 years from its date of entry into force, and unless any Contracting Party during that period requests the termination of the Agreement at the end of that period. it shall continue in force thereafter.
6. On the request addressed to the Depositary Government by any of the Governments referred to in paragraph I of this Article. consultations shall be conducted with a view to convening a meeting of representatives of the five Governments to consider the revision or amendment of this Agreement.
7. Any Party may denounce this Agreement by written notification to the Depositary Government at any time after 5 years from the date of entry into force of this Agreement. The denunciation shall take effect twelve months after the Depositary Government has received the notification.
8. The Depositary Government shall notify the Governments referred to in paragraph 1 of this Article of the deposit of instruments of ratification, approval or accession, of the entry into force of this Agreement and of the receipt of notifications of denunciation and any other communications from a Contracting Part specifically provided for in this Agreement.
9. The original of this Agreement shall be deposited with the Government of Norway which shall deliver certified copies thereof to each of the Governments referred to in paragraph I of this Article.

The Depositary Government shall transmit certified copies of this Agreement to the Secretary-General of the United Nations for registration and publication in accordance with Article 102 of the Charter of the United Nations.

In Witness Whereof the undersigned, being duly authorized by their Governments, have signed this Agreement.

Done at Oslo, in the English and Russian languages, each text being equally authentic, this 15th day of November, 1973.

Regulation 1007/2009 on Trade in Seal Products, 2009 (As amended by Regulation (EU) 2015/1775 of 2015)

The European Parliament and the council of the European Union
Having regard to the Treaty establishing the European Community, and in particular Article 95 thereof,

Having regard to the proposal from the Commission,

Having regard to the opinion of the European Economic and Social Committee,

Acting in accordance with the procedure laid down in Article 251 of the Treaty,

Whereas:

(1) Seals are sentient beings that can experience pain, distress, fear and other forms of suffering. In its declaration on banning seal products in the European Union, the European Parliament requested the Commission immediately to draft a regulation to ban the import, export and sale of all harp and hooded seal products. In its resolution of 12 October 2006 on a Community Action Plan on the Protection and Welfare of Animals 2006–2010, the European Parliament called on the Commission to propose a total import ban on seal products. In its Recommendation 1776 (2006) of 17 November 2006 on seal hunting, the Parliamentary Assembly of the Council of Europe recommended inviting the Member States of the Council of Europe practising seal hunting to ban all cruel hunting methods which do not guarantee the instantaneous death, without suffering, of the animals, to prohibit the stunning of animals with instruments such as hakapiks, bludgeons and guns, and to promote initiatives aimed at prohibiting trade in seal products.

(2) The import into Member States for commercial purposes of skins of harp seal pups and hooded seal pups and products derived therefrom is prohibited under Council Directive 83/129/EEC of 28 March 1983 concerning the importation into Member States of skins of certain seal pups and products derived therefrom

(3) Seals are hunted within and outside the Community and used for obtaining products and articles, such as meat, oil, blubber, organs, fur skins and articles made therefrom, which include products as diverse as Omega-3 capsules and garments incorporating processed seal skins and fur. Those

N. Sellheim, *International Marine Mammal Law*,
https://doi.org/10.1007/978-3-030-35268-4

products are sold commercially on different markets, including the Community market. Given the nature of those products, it is difficult or impossible for consumers to distinguish them from similar products not derived from seals.

(4) The hunting of seals has led to expressions of serious concerns by members of the public and governments sensitive to animal welfare considerations due to the pain, distress, fear and other forms of suffering which the killing and skinning of seals, as they are most frequently performed, cause to those animals.

(5) In response to concerns of citizens and consumers about the animal welfare aspects of the killing and skinning of seals and the possible presence on the market of products obtained from animals killed and skinned in a way that causes pain, distress, fear and other forms of suffering, several Member States have adopted or intend to adopt legislation regulating trade in seal products by prohibiting the import and production of such products, while no restrictions are placed on trade in these products in other Member States.

(6) There are therefore differences between national provisions governing the trade, import, production and marketing of seal products. Those differences adversely affect the operation of the internal market in products which contain or may contain seal products, and constitute barriers to trade in such products.

(7) The existence of such diverse provisions may further discourage consumers from buying products not made from seals, but which may not be easily distinguishable from similar goods made from seals, or products which may include elements or ingredients obtained from seals without this being clearly recognisable, such as furs, Omega-3 capsules and oils and leather goods.

(8) The measures provided for in this Regulation should therefore harmonise the rules across the Community as regards commercial activities concerning seal products, and thereby prevent the disturbance of the internal market in the products concerned, including products equivalent to, or substitutable, for seal products.

(9) In accordance with the Protocol on protection and welfare of animals annexed to the Treaty, the Community is to pay full regard to the welfare requirements of animals when formulating and implementing, inter alia, its internal market policy. The harmonised rules provided for in this Regulation should accordingly take fully into account considerations of the welfare of animals.

(10) To eliminate the present fragmentation of the internal market, it is necessary to provide for harmonised rules while taking into account animal welfare considerations. In order to counter barriers to the free movement of products concerned in an effective and proportionate fashion, the placing on the market of seal products should, as a general rule, not be allowed in order to restore consumer confidence while, at the same time, ensuring that animal welfare concerns are fully met. Since the concerns of citizens and consumers

extend to the killing and skinning of seals as such, it is also necessary to take action to reduce the demand leading to the marketing of seal products and, hence, the economic demand driving the commercial hunting of seals. In order to ensure effective enforcement, the harmonised rules should be enforced at the time or point of import for imported products.

(11) Although it might be possible to kill and skin seals in such a way as to avoid unnecessary pain, distress, fear or other forms of suffering, given the conditions in which seal hunting occurs, consistent verification and control of hunters' compliance with animal welfare requirements is not feasible in practice or, at least, is very difficult to achieve in an effective way, as concluded by the European Food Safety Authority on 6 December 2007.

(12) It is also clear that other forms of harmonised rules, such as labelling requirements, would not achieve the same result. Additionally, requiring manufacturers, distributors or retailers to label products that derive wholly or partially from seals would impose a significant burden on those economic operators, and would also be disproportionately costly in cases where seal products represent only a minor part of the product concerned. Conversely, the measures contained in this Regulation will be easier to comply with, whilst also reassuring consumers.

(13) In order to ensure that the harmonised rules provided for in this Regulation are fully effective, those rules should apply not only to seal products originating from the Community, but also to those introduced into the Community from third countries.

(14) The fundamental economic and social interests of Inuit communities engaged in the hunting of seals as a means to ensure their subsistence should not be adversely affected. The hunt is an integral part of the culture and identity of the members of the Inuit society, and as such is recognised by the United Nations Declaration on the Rights of Indigenous Peoples. Therefore, the placing on the market of seal products which result from hunts traditionally conducted by Inuit and other indigenous communities and which contribute to their subsistence should be allowed.

(15) This Regulation establishes harmonised rules concerning the placing on the market of seal products. It is therefore without prejudice to other Community or national rules regulating the hunting of seals.

(16) The measures necessary for the implementation of this Regulation should be adopted in accordance with Council Decision 1999/468/EC of 28 June 1999 laying down the procedures for the exercise of implementing powers conferred on the Commission.

(17) In particular, the Commission should be empowered to define the conditions for the placing on the market of seal products which result from hunts traditionally conducted by Inuit and other indigenous communities and contribute to their subsistence; to define the conditions for the import of seal products which is of an occasional nature and consists exclusively of goods for the personal use of travellers or their families; and to define the conditions for the placing on the market of seal products resulting from

hunts regulated by national law with the sole purpose of the sustainable management of marine resources. Since those measures are of general scope and are designed to amend non-essential elements of this Regulation by supplementing it with new non-essential elements, they must be adopted in accordance with the regulatory procedure with scrutiny provided for in Article 5a of Decision 1999/468/EC.

(18) With the aim of facilitating enforcement operations carried out by the relevant national authorities, the Commission should issue technical guidance notes providing non-binding indications about the codes of the Combined Nomenclature which may cover seal products subject to this Regulation.

(19) Member States should lay down rules on penalties applicable to infringements of the provisions of this Regulation and ensure that they are implemented. Those penalties should be effective, proportionate and dissuasive.

(20) Member States should report on a regular basis on the actions taken to implement this Regulation. On the basis of those reports, the Commission should report to the European Parliament and to the Council on the implementation of this Regulation.

(21) Since the objective of this Regulation, namely the elimination of obstacles to the functioning of the internal market by harmonising national bans concerning the trade in seal products at Community level, cannot be sufficiently achieved by the Member States and can therefore be better achieved at Community level, the Community may adopt measures, in accordance with the principle of subsidiarity as set out in Article 5 of the Treaty. In accordance with the principle of proportionality, as set out in that Article, this Regulation does not go beyond what is necessary in order to achieve that objective,

HAVE ADOPTED THIS REGULATION:

Article 1
Subject matter

This Regulation establishes harmonised rules concerning the placing on the market of seal products.

Article 2
Definitions

For the purposes of this Regulation, the following definitions shall apply:

1. 'seal' means specimens of all species of pinnipeds (*Phocidae*, *Otariidae* and *Odobenidae*);
2. 'seal product' means all products, either processed or unprocessed, deriving or obtained from seals, including meat, oil, blubber, organs, raw fur skins and fur

skins, tanned or dressed, including fur skins assembled in plates, crosses and similar forms, and articles made from fur skins;

3. 'placing on the market' means introducing onto the Community market, thereby making available to third parties, in exchange for payment;
4. 'Inuit' means indigenous members of the Inuit homeland, namely those arctic and subarctic areas where, presently or traditionally, Inuit have aboriginal rights and interests, recognised by Inuit as being members of their people and includes Inupiat, Yupik (Alaska), Inuit, Inuvialuit (Canada), Kalaallit (Greenland) and Yupik (Russia);

   4a. 'other indigenous communities' means communities in independent countries who are regarded as indigenous on account of their descent from the populations which inhabited the country, or a geographical region to which the country belongs, at the time of conquest or colonisation or the establishment of present State boundaries and who, irrespective of their legal status, retain some or all of their own social, economic, cultural and political institutions;
5. 'import' means any entry of goods into the customs territory of the Community.

Article 3

Conditions for placing on the market

1. The placing on the market of seal products shall be allowed only where the seal products result from hunts conducted by Inuit or other indigenous communities, provided that all of the following conditions are fulfilled:
   (a) the hunt has traditionally been conducted by the community;
   (b) the hunt is conducted for and contributes to the subsistence of the community, including in order to provide food and income to support life and sustainable livelihood, and is not conducted primarily for commercial reasons;
   (c) the hunt is conducted in a manner which has due regard to animal welfare, taking into consideration the way of life of the community and the subsistence purpose of the hunt.

   The conditions set out in the first subparagraph shall apply at the time or point of import for imported seal products.

   1a. At the time of its being placed on the market, a seal product shall be accompanied by a document attesting compliance with the conditions set out in paragraph 1 ('attesting document').

   An attesting document shall, upon request, be issued by a body recognised for that purpose by the Commission.

   Such recognised bodies shall be independent, competent to carry out their functions and subject to an external audit.
2. By way of derogation from paragraph 1, the import of seal products shall also be allowed where it is of an occasional nature and consists exclusively of goods for the personal use of travellers or their families. The nature and

quantity of those goods shall not be such as to indicate that they are being imported for commercial reasons.

3. The application of paragraphs 1 and 2 shall not undermine the achievement of the objective of this Regulation.
4. The Commission shall adopt implementing acts to further specify the administrative arrangements for the recognition of bodies that may attest to the compliance with the conditions set out in paragraph 1 of this Article and for the issuance and control of attesting documents, as well as the administrative provisions necessary for ensuring compliance with paragraph 2 of this Article. Those implementing acts shall be adopted in accordance with the examination procedure referred to in Article 5(2).
5. If there is evidence that a seal hunt is conducted primarily for commercial reasons, the Commission shall be empowered to adopt delegated acts in accordance with Article 4a in order to prohibit the placing on the market or limit the quantity that may be placed on the market of seal products resulting from the hunt concerned. It is of particular importance that the Commission follow its usual practice and carry out consultations with experts, including Member States' experts, before adopting those delegated acts.
6. The Commission shall adopt implementing acts to issue technical guidance notes setting out an indicative list of the codes of the Combined Nomenclature which may cover seal products subject to this Article. Those implementing acts shall be adopted in accordance with the examination procedure referred to in Article 5(2).

Article 4

Free movement

Member States shall not impede the placing on the market of seal products which comply with this Regulation.

Article 4a

Exercise of the delegation

1. The power to adopt delegated acts is conferred on the Commission subject to the conditions laid down in this Article.
2. The power to adopt delegated acts referred to in Article 3(5) shall be conferred on the Commission for a period of 5 years from 10 October 2015. The Commission shall draw up a report in respect of the delegation of power not later than nine months before the end of the 5-year period. The delegation of power shall be tacitly extended for periods of an identical duration, unless the European Parliament or the Council opposes such extension not later than 3 months before the end of each period.
3. The delegation of power referred to in Article 3(5) may be revoked at any time by the European Parliament or by the Council. A decision to revoke shall put an end to the delegation of the power specified in that decision. It shall take effect the day following the publication of the decision in the *Official Journal*

of the European Union or at a later date specified therein. It shall not affect the validity of any delegated acts already in force.

4. As soon as it adopts a delegated act, the Commission shall notify it simultaneously to the European Parliament and to the Council.
5. A delegated act adopted pursuant to Article 3(5) shall enter into force only if no objection has been expressed either by the European Parliament or the Council within a period of 2 months of notification of that act to the European Parliament and the Council or if, before the expiry of that period, the European Parliament and the Council have both informed the Commission that they will not object. That period shall be extended by 2 months at the initiative of the European Parliament or of the Council.

Article 5

Committee procedure

1. The Commission shall be assisted by the committee established under Article 18(1) of Council Regulation (EC) No 338/97 of 9 December 1996 on the protection of species of wild fauna and flora by regulating trade therein. That committee may call upon other existing regulatory committees as necessary, such as the Standing Committee on the Food Chain and Animal Health established by Article 58(1) of Regulation (EC) No 178/2002 of the European Parliament and of the Council of 28 January 2002 laying down the general principles and requirements of food law, establishing the European Food Safety Authority and laying down procedures in matters of food safety
2. Where reference is made to this paragraph, Article 5 of Regulation (EU) No 182/2011 shall apply.
3. As regards implementing acts to be adopted pursuant to Article 3(4), where the committee delivers no opinion, the Commission shall not adopt the draft implementing act and the third subparagraph of Article 5(4) of Regulation (EU) No 182/2011 shall apply.

Article 5a

Information

The Commission shall inform the public, with a view to raising their awareness, and competent authorities, including customs authorities, of the provisions of this Regulation and of the rules under which seal products resulting from hunts conducted by Inuit or other indigenous communities can be placed on the market.

Article 6
Penalties and enforcement

Member States shall lay down the rules on penalties applicable to infringements of this Regulation and shall take all measures necessary to ensure that they are implemented. The penalties provided for shall be effective, proportionate and dissuasive. Member States shall notify the Commission of those provisions by 20 August 2010, and shall notify it without delay of any subsequent amendment thereto.

Article 7
Reporting

1. By 31 December 2018 and every 4 years thereafter, Member States shall submit to the Commission a report outlining the actions taken to implement this Regulation.
2. The Commission shall submit to the European Parliament and to the Council a report on the implementation of this Regulation within 12 months of the end of each reporting period referred to in paragraph 1. The first report shall be submitted by 31 December 2019.
3. In its reports submitted in accordance with paragraph 2, the Commission shall assess the functioning, effectiveness and impact of this Regulation in achieving its objective.

Article 8
Entry into force and application

This Regulation shall enter into force on the 20th day following its publication in the *Official Journal of the European Union*.

Article 3 shall apply from 20 August 2010.

This Regulation shall be binding in its entirety and directly applicable in all Member States.

Index

N. Sellheim, *International Marine Mammal Law*,
https://doi.org/10.1007/978-3-030-35268-4

The manufacturer's authorised representative in the EU is Springer Nature Customer Service Centre GmbH, Europaplatz 3, 69115 Heidelberg, Germany. If you have any concerns regarding our products, please contact ProductSafety@springernature.com

Printed and bound by CPI Group (UK) Ltd, Croydon, CR0 4YY
25/09/2026
02222923-0001